Fluctuations in Physical Systems

This book provides an introduction to applied statistical mechanics by considering physically realistic models and examples. It gives a simple and accessible introduction to theories of thermal fluctuations and diffusion, and goes on to apply them in a variety of physical contexts.

The first part of the book is devoted to processes in thermal equilibrium, and considers linear systems. Starting with an introduction to the ideas underlying the subject, the text then develops the necessary mathematical tools and applies them to fluctuations and diffusion processes in a number of contexts. Ideas central to the subject, such as the fluctuation–dissipation theorem, Fokker–Planck equations and the Kramers–Kronig relations, are introduced in a natural way during the course of the exposition. The book goes on to expand the scope by including nonequilibrium systems and also illustrates some simple nonlinear systems.

This book will be of interest to final year undergraduate and graduate students studying statistical mechanics, plasma physics, basic electronics, solid-state physics and anyone who wants an accessible introduction to the subject.

HANS PÉCSELI obtained his PhD on experimental studies of plasma waves in 1974 at the Technical University of Denmark. He was research associate at the Risø National Laboratory in Denmark (1974–1991), apart from a sabbatical year at the University of Iowa, Iowa City, in 1979–1980.

He has been adjoined professor in plasma physics at the University of Tromsø, Norway, since 1989, and full professor at the Department of Physics, University of Oslo, Norway, since 1991. He has been organizer or co-organizer of several conferences and workshops, and served in a number of committees, particularly in the Scandinavian countries.

Hans Pécseli received the annual physics prize of the Danish Physical Society for 1989 together with O. Kofoed-Hansen, for works on plasma turbulence. He is a fellow of the American Physical Society, a member of the Norwegian Academy of Sciences, and a member of the Royal Danish Academy of Sciences and Letters.

He is the author or co-author of more than 150 scientific publications on plasma physics and related topics, not to mention a similar number of conference contributions.

Fluctuations in Physical Systems

HANS L. PÉCSELI

University of Oslo

CAMBRIDGE
UNIVERSITY PRESS

PUBLISHED BY THE PRESS SYNDICATE OF THE UNIVERSITY OF CAMBRIDGE
The Pitt Building, Trumpington Street, Cambridge, United Kingdom

CAMBRIDGE UNIVERSITY PRESS
The Edinburgh Building, Cambridge CB2 2RU, UK http://www.cup.cam.ac.uk
40 West 20th Street, New York, NY 10011-4211, USA http://www.cup.org
10 Stamford Road, Oakleigh, Melbourne 3166, Australia
Ruiz do Alarcón 13, 28014 Madrid, Spain

First published 2000

Printed in the United Kingdom at the University Press, Cambridge

Typeface 9½/13pt Times New Roman

A catalogue record for this book is available from the British Library

Library of Congress Cataloguing in Publication data
Pécseli, Hans L., 1947–
    Fluctuations in physical systems/Hans L. Pécseli.
        p.   cm.
    Includes bibliographical references and index.
    ISBN 0 521 65192 1 – ISBN 0 521 65592 7 (pbk.)
    1. Fluctuations (Physics) I. Title.

QC6.4.F58 P43 2000
530–dc21   99-047850

ISBN 0 521 65192 1 hardback
ISBN 0 521 65592 7 paperback

The principle of generating small amounts of *finite* improbability by simply hooking the logic circuits of a Bambleweeny 57 Sub-Meson Brain to an atomic vector plotter suspended in a strong Brownian Motion producer (such as a hot cup of tea) were of course well understood – and such generators were often used to break the ice at parties by making all the molecules in the hostess's undergarments leap simultaneously one foot to the left, in accordance with the Theory of Indeterminacy.

Many respectable physicists said that they weren't going to stand for this, partly because it was a debasement of science, but mostly because they didn't get invited to those sorts of parties.

(Douglas Adams, *The Hitchhiker's Guide to the Galaxy*)

To Henriette, Maria and Thomas

Contents

Preface

These notes were originally prepared for a series of lectures on fluctuation phenomena at the Danish Space Research Institute many years ago. In their present form, they were compiled for lectures at the University of Tromsø. I should like to thank my students and colleagues for their enthusiasm and for many stimulating discussions. In particular, I should like to express my gratitude to Professor Asger Nielsen from the Technical University of Denmark who first introduced me to this subject when I followed his lectures for my Ph.D. The present summary is indebted, in content as well as form of presentation, to his inspiring lectures and lecture notes. I am greatly indebted also to Liv Larssen at the Auroral Observatory of the University of Tromsø for her tireless and expert assistance with typing and editing the manuscript, and with preparation of figures. Also the friendly advice and support of Jan Trulsen and Bård Krane are gratefully acknowledged. The text was prepared by using LATEX and some of the figures were prepared by using Mathematica.

It is hoped that no particular prior knowledge apart from that gained from basic mathematics and statistics courses will be required. Introductory courses in thermodynamics and statistical mechanics will be a great advantage, though. Since many of the examples are taken from electronics, some basic knowledge of elementary electric circuits will also be a help. The notes advocate learning by working through examples, although this was not the original intention. On the other hand, it might not be such a bad idea after all. A detailed discussion of $1/f$ noise is deliberately omitted; this is an interesting and important topic with new developments, but it was felt to be outside the scope of the present book.

There are several repetitions in the text; it is hoped that this makes it easier to read. Care is taken to ensure that the contents of the individual chapters are interrelated, at least to some degree, and that they illustrate different aspects and consequences of some basic phenomena. There are examples and exercises in the text, where many of the examples can be considered as 'solved problems.' More can be found in the compilations of for instance Takács (1960) and Sveshnikov (1978). In particular, problems with solutions of the diffusion equation in various geometries and with various boundary conditions are discussed in great detail by Carslaw and Jaeger (1959).

The reference list has an emphasis on the classical literature on the subject and is far from complete, but ought to serve as a useful starting point for further study. In some cases a reference may seem redundant; probably nobody will check Brown's original paper on the motion named after him! On the other hand, it may be interesting to have the actual year and title of the publication of his discovery stated explicitly. Other papers, like Einstein's for instance, are certainly still worthwhile reading and are therefore given with references that are readily available, rather than the originals. Even more so with the references to Rice's papers; few have ever seen the originals in the Bell Lab. Journals, and the excellent book of reprints on statistical physics compiled by Nelson Wax is the only readily available reference for this and several other papers of fundamental importance.

An outline of individual chapters

- **Chapter 1:** This is a brief introduction containing some general discussion.
- **Chapter 2:** An outline of basics of statistical analysis is presented. The reader will, I assume, be familiar with most of the contents of this chapter, but it is considered an advantage that the text is self-contained to some degree. Moment-generating functions and characteristic functions are defined, since they will be used in later chapters. Because of their importance, the Wiener–Khinchine relations are discussed in some detail.
- **Chapter 3:** Thermal fluctuations in electric circuits are illustrated by a specific example. The model is physically realistic and deterministic, apart from a specific statistical assumption concerning a collisional process. The results illustrate the Nyquist fluctuation theorem.
- **Chapter 4:** The Nyquist fluctuation theorem, or the more general fluctuation–dissipation theorem, is derived from basic thermodynamic arguments for one specific system, and straightforward generalizations of the theorem are argued.
- **Chapter 5:** The Kramers–Kronig relations are derived and their consequences for response functions, such as dielectric functions, are discussed.
- **Chapter 6:** Brownian motion, being historically the first thoroughly described observation of thermal motion, is discussed. Simple models, as well as analytic descriptions of the phenomena, based on the Langevin equation, are presented.
- **Chapter 7:** The description of Brownian motion is completed by a simple random-walk model that can be analyzed in detail, with particular attention to the presence of reflecting or absorbing boundaries. The Lévi-flight model of a random walk is outlined.
- **Chapter 8:** Density fluctuations in gases are discussed, with emphasis on the fluctuation at a critical point.
- **Chapter 9:** A simple statistical model for random processes proposed, for instance, by Rice is discussed in detail, following his exposition. The model is used to illustrate the effect of finite record lengths on the uncertainty in estimators, for instance. Shot noise is used as an example for a practical application of the model.
- **Chapter 10:** Markov processes are discussed. Master equations are introduced, and Fokker–Planck equations derived as a special case. The chapter provides a mathematical basis for analyzing a number of stochastic or random phenomena, which is to be applied in following chapters.
- **Chapter 11:** Strictly speaking, the Fokker–Planck equation is derived for probability densities. For physical systems, an equation for particle densities in a given realization, which is not the same thing, is usually desired. In this chapter equivalents of Fokker–Planck equations are derived and applied to important problems such as sedimentation in a gravitational field and coagulation in colloids. The latter

problem is particularly interesting, in that it concerns a system that is not stationary, since the density of the macroscopic particles is changing with time.

- **Chapter 12:** Thermal fluctuations in a nonlinear circuit element are discussed, using
  a diode as an example. A general master equation is derived for the problem, and its
  solution in terms of characteristic functions demonstrated. A Fokker–Planck equation is derived as a limiting case.
- **Chapter 13:** Acceleration of light particles by random motion, Fermi acceleration, is
  discussed and illustrated by specific models. A master equation is derived and
  solved for a specific case. The analysis of the stochastic model is presented in
  terms of generating functions. A deterministic model is introduced and the possibility of chaotic phenomena demonstrated.
- **Appendixes:** Three of the six appendixes present basic statistical distributions: The
  Binomial distribution, for discrete and finite variable space, the Poisson distribution
  for discrete infinite variable space, and the Gaussian distribution for a continuous,
  infinite range of variables. Because of the frequent use of Dirac's δ-function in this
  book it was found advantageous to include also an appendix concerning its properties, without going into detail in discussions of generalized functions, though.

The letter T denotes 'temperature' throughout and always appears together with
Boltzmann's constant κ, whereas $\mathcal{T}$ indicates a (large) time interval. There being a shortage
of letters, implies that some other symbols can have different meanings in different chapters, the
worst problem being that the letter i can mean electric current as well as $\sqrt{-1}$. The correct
interpretation will in all cases be evident from the context.

1   Introduction

Many important everyday phenomena in nature appear to be unpredictable or random. Fluctuations in the neutral winds in the atmosphere are representative examples of large-scale phenomena. Examples on small scales are fluctuations in electric circuits, and random movements of tiny grains of pollen suspended in liquid; so-called Brownian motions (MacDonald, 1962). This and a number of related problems and phenomena will be discussed in some detail in the following chapters.

Although the concept of randomness may seem intuitively clear, its actual definition is somewhat ambiguous. Randomness is often associated with unpredictability, but the fact that one observer is unable to predict or comprehend a certain sequence of signals or events does not preclude the possibility that it seems perfectly transparent to another who has more *a priori* information available.[1] As an illustration of this point, consider for instance the sequence of numbers

1111111111111111, 10000, 121, 100, 31, 24, 22, ◯, 17, 16, 15, 14, 13, 12, 11, 10, . . .

and predict the number at the position indicated by ◯. Even though the sequence has been ordered in some sense and does not appear random, it does not, on the other hand, seem to be regular in any way (in particular not when we are told that the next symbol in the sequence is *G*, and so are actually also all the following ones!). However, with the proper *a priori* information the entire sequence is perfectly meaningful, and actually quite simple to comprehend.

A time-varying signal is deterministic, i.e. completely predictable, provided that it is infinitely many times differentiable and known *a priori* in a small time interval. Then, by a Taylor expansion, it can in principle be described with arbitrary accuracy to arbitrary later times. Unpredictability is therefore here related to discontinuities either in the functional values or in time derivatives at some order invalidating this expansion. Quite formally this is so, and it agrees also with intuitive expectations of random functions looking ragged. In reality it is not possible to determine all derivatives with the desired accuracy on the basis of a given time sequence and the ideal prediction outlined here is not feasible.

For a wide class of physical systems we have to accept that information can be available only on a certain level, even in the classical limit. Even though we assume that the forces acting on the molecular level are exactly known, the system may nevertheless have so many degrees of freedom that it is even in principle impossible to obtain all the relevant information on the initial conditions that would be required in order to solve the dynamic equations for the entire system. Thus we believe that we know and understand the forces acting between atoms and

---

[1] The literature on music, in particular, contains plenty of jokes on this observation. As a chairman once said at a conference on signal analysis, 'one man's signal is another man's noise.'

molecules; within classical mechanics it should be possible to describe the dynamics of, say, a cubic centimeter of gas to any degree of accuracy. However, this task would require that the initial positions of some 10^{17} atoms or molecules be known. We can safely assume that this is impossible, necessitating a statistical, or probabilistic, approach to the problem.

Even for systems with relatively *few* degrees of freedom, a somewhat similar situation may be encountered. In practice relevant initial conditions can be obtained only with a certain accuracy. The description of a wide class of physically interesting problems turns out to be extremely sensitive to the initial conditions, and the predicted temporal evolution is dramatically modified by even minute changes in these conditions. The resulting dynamic evolution can be described in statistical terms.

These few examples hint that an interpretation of statistical probabilities is that they represent systems regarding which we have insufficient *a priori* knowledge. For instance it can be argued that statistical mechanics assigns equal probabilities to all states with the same energy irrespective of the mechanical microstate simply because we do not have, and in practice *can* not have, information on these microstates. This principle of 'insufficient knowledge' can serve as a working hypothesis. It was apparently first formulated by Thomas Bayes (1763) stating that the absence of *a priori* knowledge can be expressed as *a priori* equal probabilities of events (Lee, 1989). This approach fails, however, in many respects, regarding quantum statistics in particular (Tolman, 1938, van Kampen, 1981). Mathematics can only derive the probabilities of outcomes of experiments or trials from *a priori given* probability densities or distributions. The only restrictions imposed on these probabilities are that they have to be positive definite and normalizable (and even this condition can formally be relaxed somewhat). There is no mathematical requirement that averages should exist, although it is sometimes hard to imagine describing a physical process without them.

- **Example:** The probability density

$$p_n(x) = n \frac{e^{-x}}{x} I_n(x),$$

I_n being the modified Bessel function, is normalized for $0 < x < \infty$, i.e., $\int_0^\infty p_n(x)\,dx = 1$, for all $n = 1, 2, 3, \ldots$, but it has no average, i.e. $\int_0^\infty x p_n(x)\,dx$ diverges for any value of n.

In some cases symmetry arguments can help to determine the actual probability distributions, i.e. for a die we usually assume that the probability of each face coming up is $\frac{1}{6}$. In reality even this simple case relies on an idealization in terms of an absolutely exact cube with rounded corners. Even an honest die can strictly speaking never live up to this expectation, and the assumed probability is only an approximation. For slightly more complicated situations even our intuitive understanding of symmetry arguments is not quite straightforward, as the often-quoted example of Bertrand (1889) so elegantly demonstrates. He considered the problem of a straight line drawn at random to intersect a circle with unit radius, see Fig. 1.1. The question is then this: What is the probability, P, that the chord has a length longer than $\sqrt{3}$? The answer can be argued in three different ways, unfortunately giving three different results!

(1)   Take a fixed point on the circle and consider all lines through this point assuming that the angle, θ, with the tangent is uniformly distributed in the interval $\{0; \pi\}$. All

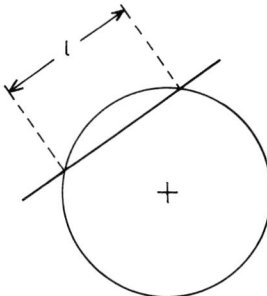

Figure 1.1 An illustration of Bertrand's problem with a circle with unit radius, where a chord is drawn at random. The chord length l is defined as the distance between the two points where the line intersects the circle.

lines cross the circle at two points except for the tangent itself, i.e. $\theta = 0$ and $\theta = \pi$, which on the other hand has a relative measure zero. In order for the chord to be longer than $\sqrt{3}$ it is required that $\pi/3 < \theta < 2\pi/3$. The interval is $\pi/3$, covering $\frac{1}{3}$ of the available interval of π, hence the answer is $P = \frac{1}{3}$.

(2) Consider all lines perpendicular to a fixed diameter of the circle. The chord is longer than $\sqrt{3}$ when the point of intersection lies on the middle half of the diameter. Consequently one finds $P = \frac{1}{2}$ by assuming the points to be uniformly distributed.

(3) For the chord to be longer than $\sqrt{3}$ its center must lie at a distance less than $\frac{1}{2}$ from the center. The area of a circle with radius $\frac{1}{2}$ is $\frac{1}{4}$ of the original circle. Assuming the chord centers to be uniformly distributed over the circle, the result $P = \frac{1}{4}$ is obtained.

The three examples are all based on the *a priori* assumption of uniform distributions, but of different quantities. Bertrand's question *has* an analytic answer, but only when it has been unambiguously formulated.

There is no obvious rescue for this and similar practical problems. There is no general method for obtaining unambiguous probability distributions for physical systems from first principles. In practice one can construct a hypothesis on the lowest possible level of description, follow its consequences for the statistical properties of the system, and eventually test the validity of the hypothesis against measurable quantities. A number of examples will illustrate this in the following chapters.

- **Exercise:** The following problem has little relevance for the present treatise, apart from giving an exercise in statistical reasoning. It is hoped that it can nevertheless give the reader some amusement before entering the more serious stuff!

  The problem originates, at least in its present form, from the journal of the Danish Engineers Society, *Ingeniøren*. It goes as follows. Four gamblers, Colt, Browning, Smith, and Wesson reach a disagreement concerning a game of poker. The problem becomes so serious that it has to be settled by a gunfight. The four gentlemen meet at sunrise and agree that they should take turns to fire one shot at a time, to be continued until there is only one survivor. There seems after all to be some sense of fairness among them, so they decide that Mr Colt who is the best shot should come last as number four, since he hits every time he shoots. Mr Browning has a hit rate of 75%, so he comes as number three, Mr Smith has a record of hitting with 50% of his shots, so he is number two, while poor Mr Wesson, with a record of only 25%, shoots first. What is Mr Wesson's optimum strategy, and what is the probability of survival for each of the four gentlemen, assuming that each follows his optimum strategy?

Understandably, the winner decides to celebrate by throwing a big party. He has now many friends, as winners have, and invites 111 guests. To have a party in style, all the seats are labeled with the guests' names. Unfortunately, the first one to arrive does not notice, and takes a seat at random (i.e. he or she *might* take the correct seat). All the others take their own seats when they arrive, provided that they are available; otherwise they take seats at random. What is the probability that the last guest to arrive ends up in the seat with the correct label? Does the answer depend in any significant way on the number of seats being even or odd?

2   Elements of statistical analysis

Assume that we have a stochastic variable defined by a range of values, each associated with a certain probability. The variable is denoted X and the values it can assume x. The variable may assume discrete values only, e.g. it can be the number of particles in a volume element, each value, x_j, associated with a probability $P(x_j)$. Heuristically, we might interpret the probability of an event as the number of 'desired' events divided by the total number of relevant events. For simple cases like honest dies and card games, this interpretation is quite adequate, but it is insufficient as a general definition.

The sample space for relevant events may be discrete and finite as in Appendix A, or discrete and *infinite* as in Appendix B. The variable can be continuous such as, e.g., in Appendix C, for instance the voltage output of a noisy amplifier, with the probability of X having a value in a narrow range $x, x + dx$ being $P(x)\, dx$. Finally a combination of the two can occur, i.e. a mixture of discrete and continuous states as is encountered in atomic physics. The set of values, or *states*, that X can assume may be multidimensional, in which case it can conveniently be written as a vector $\mathbf{X}$. An example is the velocity components of a randomly moving particle.

By statistical averaging we understand the process based on the assumption that, ideally, infinitely many realizations are available. The average value of, say, X is then obtained by *ensemble averaging*:

$$\langle X \rangle = \lim_{N \to \infty} \frac{{}^1x + {}^2x + {}^3x + \cdots + {}^Nx}{N}, \tag{2.1}$$

where the indexes refer to the labels of individual realizations in the ensemble. If this is done in practice, only a finite number, N, of realizations is available, and only an *estimate*, i.e. an *approximation* to the actual average, can be achieved in this way. In terms of the normalized probability density $P(x)$, the mathematical process of averaging is expressed as

$$\langle X \rangle = \int_{-\infty}^{\infty} xP(x)\, dx, \tag{2.2}$$

for a continuous variable or, alternatively, in terms of a sum for discrete variables. It is evidently assumed that the probability density, $P(x)$, is known *a priori*. More generally, the average, or mean, of any quantity $f(X)$ is defined as $\langle f(X) \rangle = \int_{-\infty}^{\infty} f(x)P(x)\, dx$. Often we find the term *expectation value* $E\{f(X)\}$ for $\langle f(X) \rangle$. The term 'expectation value' might be somewhat misleading; assume that we have an honest die and calculate the average or expectation of the number $\mathcal{N}$ of dots. Since the numbers $\mathcal{N} = 1, 2, 3, 4, 5,$ and 6 are equally probable, we easily obtain $\langle \mathcal{N} \rangle = 3.5$. It is, however, unwise ever to actually *expect* the number 3.5 to come up, for only integer numbers can occur.

It may be important to note that, even though an event has zero probability, this does not necessarily mean that the event is physically impossible. The probability of an honest die showing 6 in *every* trial is zero, but there are no physical laws prohibiting this from actually happening.[1]

Probabilities can be expressed in terms of a *distribution function* or cumulative distribution function (Hogg and Craig, 1970), $\Pr(x)$, with the relation

$$\Pr(x) = \int_{-\infty}^{x} P(x')\,dx'$$

for the one-variable case. While $P(x)\,dx$ denotes the probability of finding x in the small interval $\{x, x + dx\}$, the distribution function $\Pr(x)$ gives the probability of finding x in the interval $\{-\infty, x\}$. For discrete variables the probability distribution is sometimes preferred, in order to avoid using the δ-functions (see Appendix D) which have to be introduced in the probability density for this case. The formulation in the following will be using only probability densities.

2.1   One-variable probabilities

First we discuss the case in which we are dealing with probabilities of one variable, say $P(x)$. Often we are not really interested in all the information contained in $P(x)$, but are content with averages such as $\langle X^n \rangle$, for some values of n. This information is more readily obtained, for instance, from moment-generating functions.

2.1.1   Generating functions

The average of $\exp(\alpha x)$, called the *moment-generating function*, is defined as (Bendat, 1958, Hogg and Craig, 1970)

$$M(\alpha) \equiv \langle \exp(\alpha X) \rangle = \int_{-\infty}^{\infty} e^{\alpha x} P(x)\,dx. \tag{2.3}$$

In particular we find by using (2.3) the derivatives

$$\frac{d^n M(\alpha)}{d\alpha^n} = \int_{-\infty}^{\infty} x^n e^{\alpha x} P(x)\,dx.$$

---

[1] It is often said that it does not make sense to discuss the probability of events that *have* already occurred, and of course much can be said in justification of this statement. However, even proponents of this point of view will probably (like the author and most readers) start wondering if the opponent in a die game continuously gets 6 or whatever number is needed to win. Faced with an event that has already happened, it is justifiable to seek its explanation by considering the probabilities of various causes for its occurrence. When different explanations for an event are possible, giving preference to the cause with the highest probability is a fully acceptable procedure. Maximum-likelihood methods (Hogg and Craig, 1970) are based on this basic hypothesis, by virtue of the assumption that it is the most probable event which is actually being observed.

The moment-generating function is evidently a function of the variable α and serves to generate the averages, or *moments*, of X, i.e. $\langle X \rangle$, $\langle X^2 \rangle$, etc. which are obtained by evaluating $dM(\alpha)/d\alpha$, $d^2 M(\alpha)/d\alpha^2$, etc., at $\alpha = 0$. The individual terms in a Taylor expansion of $M(\alpha)$ will consequently contain the averages $\langle X^n \rangle$ in increasing order. However, not every probability density has an associated moment-generating function, and it is often advantageous to consider the *characteristic function* defined as the Fourier transform of the probability density

$$CH(\alpha) \equiv \langle \exp(i\alpha X) \rangle = \int_{-\infty}^{\infty} e^{i\alpha x} P(x)\, dx. \tag{2.4}$$

With rather mild restrictions on the variation of $P(x)$ at large $|x|$, the moments of X are then given by

$$\langle X^n \rangle = (-i)^n \frac{d^n}{d\alpha^n} CH(\alpha) \Big|_{\alpha=0}, \tag{2.5}$$

the subscript indicating that the derivative is to be taken at $\alpha = 0$. For a discrete variable, the characteristic functions

$$CH(\alpha) \equiv \sum_{j}^{\infty} P(x_j) e^{i\alpha x_j} \tag{2.6}$$

can be introduced, where $P(x_j)$ is the probability of the jth event. Here, the characteristic function is a sum of exponentials.

For the case in which a variable n takes on only integer values, a definition in terms of the z-transform (Oppenheim and Schafer, 1975) gives the generating function

$$\Gamma(z) \equiv \langle Z^n \rangle = \sum_{n=-\infty}^{\infty} z^n P_n. \tag{2.7}$$

Here n is a *lattice-type* random variable. We recognize $\Gamma(1/z)$ as the z-transform of P_n with n taking on only integer values (Papoulis, 1991). (The Γ-function introduced here should not be confused with the gamma-function which interpolates n!) On differentiating (2.7) k times we obtain

$$d^k \Gamma(z)/dz^k = \langle n(n-1)\cdots(n-k+1)z^{n-k} \rangle,$$

which for $z = 1$ becomes $\langle n(n-1)\cdots(n-k+1) \rangle$.

- **Exercise:** Derive the moment-generating function for a Poisson distribution.

- **Exercise:** Obtain the moment-generating function and the characteristic function for the binomial distribution.

- **Exercise:** Derive the moment-generating function for a Gaussian distribution $P(x) = (2\pi A)^{-1/2} \exp(-\frac{1}{2} x^2/A)$ and demonstrate that $\langle X^4 \rangle = 3\langle X^2 \rangle^2$.

- **Exercise:** Demonstrate that the characteristic function for a variable Z that is the sum of two independent random variables, $Z = X + Y$, is the product of the individual characteristic functions. Generalize this result to a sum of arbitrarily many independent variables.

- **Example:** By a random vector L, we understand a vector drawn in a random direction and possibly with length, L, chosen at random also (Feller, 1971). For a vector of *unit* length in three dimensions, $\mathcal{R}^3$, the distribution of the projection, L_x, on an axis is uniform over $\{0, 1\}$. The projection of the same vector on a *plane* has a probability density $\ell/\sqrt{1-\ell^2}$ for $0 < \ell < 1$.

It is important to note that the result depends on the dimensions of the problem. For a vector of unit length in two dimensions, $\mathcal{R}^2$, the distribution of the projection, L_x, on an axis is $2/(\pi\sqrt{1-\ell^2})$ for $0 < \ell < 1$.

Consider now the sum of two independent random unit vectors in $\mathcal{R}^2$. The resultant of these vectors has a length L, with a probability density $2/(\pi\sqrt{4-\ell^2})$ for $0 < \ell < 2$. Actually, by the law of cosines, $L = \sqrt{2-2\cos\gamma} = |2\sin\left(\frac{1}{2}\gamma\right)|$, where γ is the angle between the two vectors, and $\frac{1}{2}\gamma$ is by assumption uniformly distributed in $\{0, \pi\}$.

2.1.2   Changes of variables

An important question concerns the change in variables; often we know *a priori* the probability density for a certain event, and want to determine the probability density for something else that depends in a deterministic way on the outcome of this event. As a simple example, we may consider X to be a temperature and Y to be the length of a metal rod, which is varying due to thermal expansion in some, possibly nonlinear, way. Assume that the probability density, $P_X(x)$, of the event X is known, and that another event Y is a deterministic consequence of X, i.e. $Y = f(X)$. It is then, at least in principle, straightforward to determine the probability density $P_Y(y)$. The probability that Y has a value in the range $\{y; y + \Delta y\}$ is

$$P_Y(y) = \int \delta[f(x) - y]P_X(x)\,dx. \tag{2.8}$$

In the case in which there is a one-to-one correspondence between X and Y, the transformation is readily expressed as

$$P_Y(y)\,dy = P_X(x)\,dx. \tag{2.9}$$

In case there are problems with the sign (probabilities had better be positive numbers!) this expression is to be interpreted as $P_Y(y) = P_X(x)|J|$, J being the Jacobian determinant for the general multivariable case. The expression (2.9) can be most useful.

- **Exercise:** Assume that a circle with radius R is placed randomly with its center on the x-axis, with positions uniformly distributed in the interval $\{0; \mathcal{L}\}$. Determine the probability density of chord lengths along the y-axis, i.e. the distribution of the lengths of segments of the y-axis inside the circle. Let the ratio $R/\mathcal{L}$ be arbitrary. Repeat the problem, now with a *sphere* with radius R placed randomly in the x–z plane, with its center uniformly distributed in a square $\{0; \mathcal{L}\}, \{0; \mathcal{L}\}$.

- **Exercise:** Assume that the variable X has a Gaussian distribution, $P(x) = (2\pi\sigma^2)^{-1/2}\exp(-\frac{1}{2}x^2/\sigma^2)$. Consider the variable $Y = X^2$ and demonstrate that its probability density is given by the *chi-square* probability density

$$P(y) = \frac{1}{\sigma\sqrt{2\pi y}}\exp\left(-\frac{y}{2\sigma^2}\right), \tag{2.10}$$

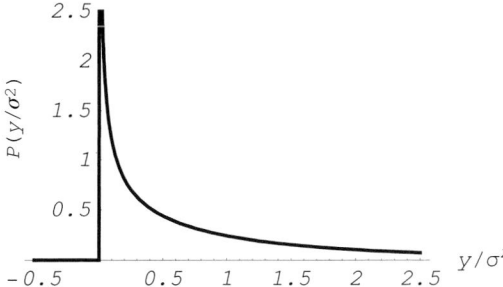

Figure 2.1 The normalized chi-square distribution, $P(y/\sigma^2)$, as given by (2.10).

for $y \geq 0$ and $P(y) = 0$ otherwise, see Fig. 2.1. Demonstrate that $\langle Y \rangle = \sigma^2$ and $\langle Y^2 \rangle = 3\sigma^4$. Prove explicitly that $P(y)$ is indeed normalized, $\int_0^\infty P(y)\,dy = 1$.

- **Exercise:** Assume that a variable X has a Gaussian distribution,

$$P(x) = (2\pi\sigma^2)^{-1/2} \exp(-\tfrac{1}{2}x^2/\sigma^2).$$

Consider another variable

$$Y = (2/\pi)^{-1/2} \int_0^X \exp(-\tfrac{1}{2}y^2)\,dy \text{ for } X \geq 0,$$

while $Y = 0$ for $X < 0$. Derive the probability density for Y.

2.2   Multivariable probabilities

The random variable can be multidimensional, as has already been mentioned, i.e. the corresponding probabilities can depend on many different variables. Averages are defined by a simple generalization of (2.2) as

$$\langle f(X_1, X_2, \ldots, X_N) \rangle = \int_{-\infty}^{\infty} f(x_1, x_2, \ldots, x_N) P(x_1, x_2, \ldots, x_N)\,dx_1\,dx_2 \ldots dx_N, \quad (2.11)$$

where $P(x_1, x_2, \ldots, x_N)$ is the joint probability density for the variables $x_1, x_2, \ldots, x_N$.

For statistically independent variables, their joint probability density is the product of their individual probability densities, e.g.

$$P(x_1, x_2, \ldots, x_n) = P(x_1)P(x_2) \ldots P(x_N).$$

Pairwise independence does not ensure absolute independence; assume for instance $P(x_1, x_2) = P(x_1)P(x_2)$ and $P(x_2, x_3) = P(x_2)P(x_3)$, but this does not mean that x_1 and x_3 are independent, i.e. $P(x_1, x_3) \neq P(x_1)P(x_3)$ in general. The proof is trivial.

Moment-generating functions and characteristic functions can be defined for multivariate distributions as well. For instance, for a bivariate probability density, $P(x_1, x_2)$, we have

$$CH(\alpha, \xi) \equiv \langle \exp(i\alpha X_1 + i\xi X_2) \rangle$$

$$= \int_{-\infty}^{\infty} e^{i\alpha x_1 + i\xi x_2} P(x_1, x_2)\,dx_1\,dx_2. \quad (2.12)$$

Averages or moments are then given by

$$\langle X_1^n X_2^m \rangle = (-i)^{n+m} \frac{d^{n+m}}{d\alpha^n d\xi^m} CH(\alpha, \xi) \Big|_{\alpha=\xi=0}. \tag{2.13}$$

A probability can be *conditional*, with $P(x|y)$ giving the probability for the event x, given with certainty the event y. Then, $P(x) = \int_{-\infty}^{\infty} P(x|y)\,dy$ for a continuous variable, or a corresponding sum for a discrete variable. By Bayes' rule we have

$$P(x|y) = \frac{P(x, y)}{P(y)}, \tag{2.14}$$

where $P(x, y)$ is the joint probability for x and y. Bayes' rule has self-evident generalizations to multivariable probabilities. It is often easier to predict or argue expressions for *conditional* probabilities than it is for their *unconditional* counterparts. The inverted version of (2.14) can then sometimes be used to obtain the full joint probability.

- **Examples:** A harmonic oscillation $\tau_1 \cos(t + \tau_2)$ with a random phase τ_2 and a random amplitude τ_1 represents a sample function with two parameters. A function that alternates between $+1$ and -1 at N random times $t = \tau_n$ with $n = 1, 2, \ldots, N$ represents a sample function with N parameters.

2.2.1   Correlation

The *covariance* of two random variables X and Y can be defined as

$$\begin{aligned} Cov &\equiv \langle (X - \langle X \rangle)(Y - \langle Y \rangle) \rangle \\ &\equiv \langle XY \rangle - \langle X \rangle \langle Y \rangle. \end{aligned} \tag{2.15}$$

The *correlation coefficient* for the relationship between two random variables X and Y can be defined as

$$C \equiv \frac{\langle (X - \langle X \rangle)(Y - \langle Y \rangle) \rangle}{\sqrt{\langle (X - \langle X \rangle)^2 \rangle \langle (Y - \langle Y \rangle)^2 \rangle}}. \tag{2.16}$$

For the case $X = Y$ we have $C = 1$, identically. If, on the other hand, X and Y are independent variables, we readily find that $C = 0$. For complex variables, we define $Cov \equiv \langle XY^* \rangle - \langle X \rangle \langle Y^* \rangle$, where the asterisk denotes the complex conjugate.

2.3   Stochastic processes

Consider a measurable quantity $Y(t)$ that is a function of some variable t. Let $Y_x(t)$ denote an ensemble of such functions of t, labeled by a random parameter x with a given distribution, $P(x)$. When t stands for time it is customary to call $Y(t)$ a random or stochastic process. A *sample function* or *a realization* of the process is obtained for one particular value of x. An ensemble of such sample functions thus constitutes a stochastic process. The averaging is understood as an ensemble averaging in the sense indicated in (2.1).

- **Examples:** A particularly simple example of a stochastic process is one in which each sample function is for all times a constant x that varies over the ensemble. A harmonic oscillation $\cos(t + x)$ with a random phase x represents a sample function with one parameter.

Let $P_1(y_1, t_1)\,dy_1$ be the probability that $Y(t)$ at time $t = t_1$ lies in the interval $\{y_1; y_1 + dy_1\}$. Then

$$P_1(y_1, t_1) = \int_{-\infty}^{\infty} \delta[Y_x(t_1) - y_1]P(x)\,dx. \tag{2.17}$$

The ensemble average of $Y(t_1)$ is then obtained by writing

$$\langle Y(t_1)\rangle = \int_{-\infty}^{\infty} y_1 P_1(y_1, t_1)\,dy_1,$$

and similarly for other averages.

Generalizing the foregoing expression, we let $P_2(y_1, t_1; y_2, t_2)\,dy_1\,dy_2$ be the joint probability that $Y(t_1)$ lies in the interval $\{y_1; y_1 + dy_1\}$ at time t_1 and $Y(t_2)$ lies in the interval $\{y_2; y_2 + dy_2\}$ at time t_2. Quite generally we have

$$P_N(y_1, t_1; y_2, t_2; \ldots; y_N, t_N) = \int_{-\infty}^{\infty} \delta[Y_x(t_1) - y_1] \cdots \delta[Y_x(t_N) - y_N]P(x)\,dx. \tag{2.18}$$

It is natural to assume a time ordering $t_n \geq t_{n-1} \geq \cdots \geq t_2 \geq t_1$, noting that some of the times can actually be equal. One can thus define a hierarchy of probabilities $P_1, P_2, P_3, \ldots$ that serves to describe the stochastic process.

Again it can be argued that the probability densities can contain too much information. For practical purposes, the lesser amount of information contained in the first few moments of the distributions will often suffice. Such moments are, for a continuous one-point probability density, defined as

$$\langle Y^n \rangle \equiv \int_{-\infty}^{\infty} y_1^n P_1(y_1)\,dy_1, \tag{2.19}$$

for $n = 1, 2, \ldots$. The average value, for $n = 1$, is often known *a priori* from symmetry arguments. The standard deviation, or root mean square, $\sigma \equiv \langle (Y - \langle Y\rangle)^2\rangle^{1/2}$ measures the level of fluctuations. The *skewness*, $S \equiv \langle (Y - \langle Y\rangle)^3\rangle/\sigma^3$, measures the lack of symmetry of the signal, while the *kurtosis*, $K \equiv \langle (Y - \langle Y\rangle)^4\rangle/\sigma^4$, is a measure of the excursion of the signal to values in excess of the standard deviation. For a Gaussian random signal we have $S = 0$ and $K = 3$. Using these values as reference ones, the kurtosis is sometimes defined as $K \equiv \langle (Y - \langle Y\rangle)^4\rangle/\sigma^4 - 3$ (Abramowitz and Stegun, 1972). Signals characterized by a large kurtosis are often 'bursty' in nature. See Fig. 2.2 for illustrative examples.

2.3.1   Correlation functions

With an ensemble of time records available for the fluctuating quantity Y, we can construct averages such as $\langle Y^n(t_1)Y^m(t_2)\rangle$ referring to different times t_1 and t_2 in the same record. Particularly important is the auto-correlation function defined as

$$R(t_1, t_2) \equiv \langle Y(t_1)Y(t_2)\rangle$$
$$= \iint_{-\infty}^{\infty} y_1(t_1)y_2(t_2)P_2(y_1, t_1; y_2, t_2)\,dy_1\,dy_2, \tag{2.20}$$

(a)

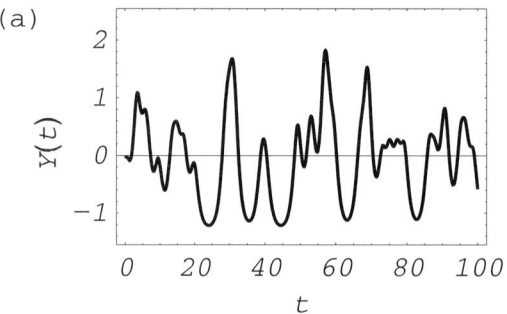

(b)

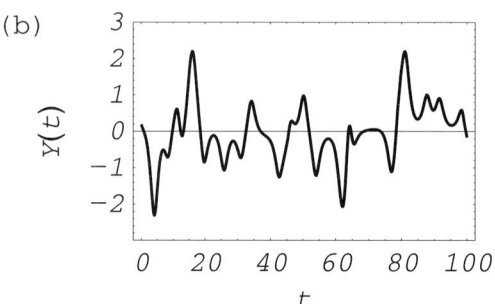

Figure 2.2 An illustration of a signal $Y(t)$ having positive skewness in (a). A signal with vanishingly small skewness but large kurtosis is illustrated in (b).

where $R(t_1, t_1) \neq R(t_2, t_2)$ in general. As has been mentioned, the function $R(t_1, t_2)$ refers to events at two different times but in the same record. This is sometimes emphasized by adding subscripts such as in $R_{YY}(t_1, t_2)$. For complex variables, the correlation function is defined as $R(t_1, t_2) \equiv \langle Y(t_1) Y(t_2)^* \rangle$, where the asterisk denotes the complex conjugate. If a time record is sampled at N points, $t_1, t_2, \ldots, t_N$, not necessarily equidistant, a correlation matrix with elements $R_{ij} \equiv R(t_i, t_j)$, where evidently $R_{ij} = R_{ji}$, can be defined.

With reference to optics (Papoulis, 1991), an autocovariance or coherence function can be defined as $R(t_1, t_2) - \langle Y(t_1) \rangle \langle Y(t_2) \rangle$. The correlation coefficient or complex degree of coherence is defined as

$$C(t_1, t_2) \equiv \frac{\langle Y(t_1) Y(t_2) \rangle - \langle Y(t_1) \rangle \langle Y(t_2) \rangle}{\sqrt{(\langle Y(t_1)^2 \rangle - \langle Y(t_1) \rangle^2)(\langle Y(t_2)^2 \rangle - \langle Y(t_2) \rangle^2)}} \tag{2.21}$$

with the property $C(t_1, t_1) = C(t_2, t_2) = 1$. The correlation coefficient is a normalized equivalent of the autocovariance (Papoulis, 1991).

Considering n different times $t_1, t_2, \ldots, t_n$ in the ensemble of records of the signal $Y(t)$, we can construct $a_1 Y(t_1) + a_2 Y(t_2) + \cdots + a_n Y(t_n)$, where a_i are the elements of an arbitrary, nonzero, n-dimensional vector $\boldsymbol{a}$. We have evidently

$$\left\langle \left(\sum_i a_i Y(t_i) \right)^2 \right\rangle \equiv \sum_i \sum_j a_i \langle Y(t_i) Y(t_j) \rangle a_j$$

$$\equiv \sum_i \sum_j a_i R(t_i, t_j) a_j \geq 0, \tag{2.22}$$

for any nontrivial real vector, $\boldsymbol{a}$. For complex $\boldsymbol{a}$, we use the complex conjugate a_j^* in (2.22). The correlation function $R(t_i, t_j)$ is said to be *non-negative definite*. This property of correlation functions will be important in discussions of their Fourier transforms. In cases in which only the inequality sign in (2.22) applies, the correlation function is said to be *positive definite*. The distinction between positive definite and non-negative definite is related to the notion of linear independence (Papoulis, 1991). The correlation function $R(t_i, t_j)$ can be interpreted as the (i, j)th element of a symmetric tensor, since evidently $R(t_i, t_j) = R(t_j, t_i)$.

- **Example:** Consider the simple matrix

$$\{M(i, j)\} = \begin{vmatrix} 1 & 1 & 0 \\ 1 & 1 & 1 \\ 0 & 1 & 1 \end{vmatrix}, \tag{2.23}$$

  and let $\{a_i\} = \{1, -1, 1\}$. It is easily demonstrated that, for this particular case, $\sum_{i,j} a_i M(i, j) a_j = -1$, i.e. it is not sufficient for a matrix to have positive elements only, in order to ensure that it be positive definite. The matrix M given by (2.23) can not represent a correlation matrix.

2.4   Stationary stochastic processes

Particularly important are *stationary* random processes (Davenport and Root, 1958, Yaglom, 1962, Cox and Miller, 1977) defined by the requirement that all P_N depend on time separations only in the sense that

$$P_N(y_1, t_1 + \tau; y_2, t_2 + \tau; \ldots; y_N, t_N + \tau) = P_N(y_1, t_1; y_2, t_2; \ldots; y_N, t_N), \tag{2.24}$$

for all N and τ. In particular $P_1(y_1)$ is independent of time. Shot noise is a particular stationary process defined by

$$P_N(y_1, t_1; y_2, t_2; \ldots; y_N, t_N) = P(y_1, t_1)^N$$

for all N, i.e. the events at $t_1, t_2, \ldots, t_N$ are independent. Hence, the statistical properties of shot noise are described by a single probability density. An alternative name is *a Poisson process*.

- **Example:** From the definition (2.24) of a stationary stochastic process we find for $N = 1$ that $P_1(y_1, t_1) = P_1(y_1, t_1 + \tau)$. This relation remains valid also for $\tau = -t_1$. Since the origin for the time axis $t = 0$ is chosen arbitrarily, we argue that $P_1(y_1, t_1) = P_1(y_1)$, independent of time. Similarly, for $N = 2$ we have $P_2(y_1, t_1; y_2, t_2) = P_2(y_1, t_1 + \tau; y_2, t_2 + \tau)$, e.g. also for $\tau = -t_1$. Consequently, $P_2(y_1, t_1; y_2, t_2) = P_2(y_1; y_2; \tau)$, with $\tau = |t_2 - t_1|$.

A stationary random process can be *ergodic*, in which case ensemble averages equal time averages, e.g. the correlation function can be written as

$$\langle Y(t_1) Y(t_1 + \tau) \rangle = \lim_{T \to \infty} \frac{1}{T} \int_0^T {}^k Y(t_1)^k Y(t_1 + \tau) \, dt_1, \tag{2.25}$$

no matter which member of the ensemble (labeled by k) is chosen. For these processes the random nature of the signal manifests itself in all detail if we wait long enough. Ergodic processes are important by virtue of the fact that they allow averages to be obtained from one (long) time series, i.e. one sample function or realization, rather than an ensemble of many such realizations (Khinchine, 1949).

- **Exercise:** Give an example of a *non*stationary stochastic process in which $P_1(y_1)$ is nevertheless independent of time.

- **Exercise:** Demonstrate by examples that stationary random processes need not be ergodic.

- **Exercise:** Obtain an explicit expression for the correlation function of a signal (a *random telegraph wave*) that alternates between $+1$ and -1 at irregular intervals, such that the probability of a change of sign in a small time interval Δt is $\nu \Delta t$, where ν is a given constant; see also Appendix B. Give the result for the case in which the signal alternates between 0 and $+1$.

Although it can be defined quite generally, the *correlation function* given by (2.20) is particularly useful for characterizing *stationary* random processes for which it is a function of the time separation $\tau \equiv t_1 - t_2$ only, $R(t_1, t_2) = R(\tau)$. A triple correlation, as well as higher order correlations, can be defined similarly. Some random processes are stationary in a *wide sense*, meaning that their average is constant and their correlation function depends on time separations only, although this need not be true for other averages such as triple correlations etc. Random processes satisfying (2.24) are *stationary in a strict sense* (Davenport and Root, 1958).

2.4.1   Cross-correlations

The foregoing discussion implicitly assumed that the variable Y refers to *one* particular quantity, say the fluctuating output from a mistuned radio receiver. Each realization of the ensemble contains one record for the time-varying quantity, Y. It is customary to denote the correlation function for this case the *auto-correlation* function. More generally we consider joint probability densities from two signals of different origins, and let x_1 and y_1 refer to different time records, say $X(t)$ and $Y(t)$, i.e. each realization of the ensemble contains *two* records, one for X and one for Y. The correlation between these two records is called the *cross-correlation* and we indicate this by using subscripts such as in R_{XY},

$$R_{XY}(t_1, t_2) \equiv \langle X(t_1) Y(t_2) \rangle$$
$$= \iint_{-\infty}^{\infty} x_1(t_1) y_1(t_2) P_2(x_1, t_1; y_1, t_2) \, dx_1 \, dy_1, \tag{2.26}$$

where $P_2(x_1, t_1; y_1, t_2) \, dx_1 \, dy_1$ is the probability that the variables X at time t_1 and Y at time t_2 assume values in the narrow intervals $x_1, x_1 + dx_1$ and $y_1, y_1 + dy_1$, respectively. The notation given before will remain unaffected but clearly we will have $R_{XY}(t_1, t_2) \neq R_{XY}(t_2, t_1)$, in general. For time-stationary processes, X and Y, the cross-correlation is a function of time separations only, just like the auto-correlation, i.e. $R_{XY}(\tau)$. The arguments are readily generalized to the joint probability density of any number, N, of different signals and their cross-correlations.

- **Exercise:** Express the correlation function in terms of the appropriate bivariate characteristic function.

If two signals X and Y are statistically independent, their joint probability density $P_2(x_1, t_1; y_1, t_2) = P(x_1, t_1)P(y_1, t_2)$ in (2.26), and consequently $R_{XY}(t_1, t_2) = \langle X(t_1) \rangle \langle Y(t_2) \rangle$ and the coherence function (2.21) vanishes. Statistically independent records are uncorrelated. It can, on the other hand, *not* be argued that uncorrelated records should be independent! It is quite easy to imagine cases in which particular forms for the joint probability $P_2(x_1, t_1; y_1, t_2)$ happen to give $R_{XY}(t_1, t_2) = 0$, at least for some selected times t_1, t_2. The following chapters will give such examples. There is, however, one special case, signals having Gaussian statistics, for which it will be demonstrated that uncorrelated events are also statistically independent. Particular attention will be paid to this example.

2.4.2   Properties of correlation functions

Some basic properties of correlation functions can readily be summarized. The auto-correlation function for time-stationary random processes is symmetric,

$$R_{XX}(\tau) = R_{XX}(-\tau).$$

For the *cross*-correlation, it is evident from the definition (2.26) that the time symmetry is lost, i.e. $R_{XY}(\tau) \neq R_{XY}(-\tau)$, in general.

Next, it can be demonstrated that, for stationary ergodic processes, the auto-correlation function assumes its maximum value at $\tau = 0$, i.e.

$$R_{XX}(0) \geq R_{XX}(\tau). \tag{2.27}$$

The proof follows from the Schwartz inequality, which is valid for arbitrary integration limits and under quite general conditions:

$$\int [f(t)]^2 dt \int [g(t)]^2 dt \geq \int [f(t)g(t)\, dt]^2,$$

with equality occurring when $f(t)/g(t)$ equals a constant (Bendat, 1958). Considering an auto-correlation as a measure of the ability of a function to interfere with itself, it can be argued that, physically, the result (2.27) means that any signal exhibits maximum interference with itself at zero time delay. One exception is a harmonic signal, $\cos(2\pi t/T)$, for which the auto-correlation function reaches the same peak value at regular times, $R_{XX}(\tau = nT) = R_{XX}(0)$, for $n = 1, 2, \ldots$, exemplifying the equality sign in (2.27).

As mentioned before, a cross-correlation function is non-negative definite. Not any symmetric function that is maximized by its value at the origin will do as an auto-correlation function. A trivial counter-example is a box function, which is constant for $|t| \leq \tau$ and vanishes elsewhere.

- **Exercise:** Prove the Schwartz inequality by observing that the integral $\int [af(t) + bg(t)]^2 dt > 0$ for all real a and b, unless $af(t) + bg(t) = 0$. A generalization of the theorem is called *Hölder's inequality*; see for instance Titchmarsh (1950).

A *cross*-correlation function can assume its maximum value anywhere. One particular case deserves a remark. Often, the signal Y is just a delayed version of the signal X, and in this case the cross-correlation is just a shifted copy of the auto-correlation. Thus, if $Y(t) = X(t + \Delta t)$, we have $R_{XY}(\tau) = R(\tau - \Delta t)$. A measurement of the cross-correlation between a signal at its emitter and at a distant receiver can thus serve to measure a time delay by measuring the shift in the maximum and thus, for instance, a speed of propagation.

2.4.2.1   Mixed spatial–temporal variables

Assume for instance that N signals are detected by obtaining a space–time-varying signal at N spatially distributed detectors located at positions r_j for $j = 1, 2, \ldots, N$. The joint probability density can then be specified as $P(r_1, r_2, \ldots, r_N; t_1, t_2, \ldots, t_M)$ with $N \neq M$ in general. Generalizing the notion of time-stationarity, we call the space–time-varying signal spatially *homogeneous* if

$$P_N(r_1, r_2, \ldots, r_N; t_1, t_2, \ldots, t_M) =$$
$$P_N(r_1 + r_0, r_2 + r_0, \ldots, r_N + r_0; t_1, t_2, \ldots, t_M)$$

for all N and all combinations of positions. The signal is spatially *isotropic* if P_N is independent of the *direction* of r_0. The correlation function, in particular, becomes $R(r_i, r_j; t_i, t_j) = R(r_i - r_j; t_i - t_j)$ for time-stationary and spatially homogeneous signals. The symmetry relation of the correlation function for time-stationary, spatially homogenous processes becomes

$$R(r_i - r_j; t_i - t_j) \equiv R(\xi_i, \tau) = R(-\xi_j, -\tau)$$

with $\xi_i = r_i - r_j$ and $\tau = t_i - t_j$. Note that a signal can, of course, be spatially homogeneous and isotropic without being time-stationary and vice versa.

2.4.3   Statistics of time derivatives

Given a set of realizations of a differentiable random process, new ones can be generated by differentiation. It might then be desired to determine, for instance,

$$\left\langle Y(t) \frac{d^n Y(t)}{dt^n} \right\rangle,$$

assuming the derivatives to exist. To analyze this expression, it is first noted that differentiation and statistical averaging commute. This can be seen by noting that differentiation can be written as a difference $dY(t)/dt \approx [Y(t + \Delta t) - Y(t)] / \Delta t$ and that the average of a sum (difference) is the sum (difference) of the individual averages. We may write, for instance,

$$\left\langle Y(t) \frac{d^n Y(t)}{dt^n} \right\rangle = \frac{d}{dt} \left\langle Y(t) \frac{d^{n-1} Y(t)}{dt^{n-1}} \right\rangle - \left\langle \frac{dY(t)}{dt} \frac{d^{n-1} Y(t)}{dt^{n-1}} \right\rangle. \tag{2.28}$$

For time-stationary random processes the first term is vanishing and by induction it is easily demonstrated that

$$\left\langle Y(t) \frac{d^n Y(t)}{dt^n} \right\rangle = (-1)^p \left\langle \frac{d^p Y(t)}{dt^p} \frac{d^{n-p} Y(t)}{dt^{n-p}} \right\rangle, \tag{2.29}$$

for all $p < n$. In particular, we find that

$$\left\langle \frac{d^{2n+1} Y(t)}{dt^{2n+1}} \frac{d^{2m} Y(t)}{dt^{2m}} \right\rangle = 0. \tag{2.30}$$

As a special case $\langle Y(t) \, dY(t)/dt \rangle = 0$ for a stationary process, i.e. a signal is uncorrelated to its time derivative taken at the same time (but of course not statistically independent, in general). The foregoing results evidently assume that all the derivatives exist *at all times*.

Define now $\langle \Delta Y^2(t) \rangle$ as the mean-square change in value over time t of a real stationary function $Y(t)$. Then

$$\begin{aligned}
\langle \Delta Y^2(t) \rangle &= \lim_{T \to \infty} \int_{-T}^{T} [Y(\tau + t) - Y(\tau)]^2 \, d\tau \\
&= \lim_{T \to \infty} \int_{-T}^{T} [Y^2(\tau + t) + Y^2(\tau) - 2Y(\tau + t) Y(\tau)] \, d\tau \\
&= 2 \langle Y^2(t) \rangle - 2R(t). \tag{2.31}
\end{aligned}$$

For sufficiently small times we may write $\langle \Delta Y^2(t) \rangle \approx \langle \Delta(Y(t_1 + t) - Y(t))^2 \rangle \approx \langle (t \, dY(t_1)/dt)^2 \rangle$ and consequently approximate

$$R(t) \approx \langle Y(t)^2 \rangle - \frac{1}{2} t^2 \left\langle \left(\frac{dY(t)}{dt} \right)^2 \right\rangle, \tag{2.32}$$

giving an important interpretation of the curvature of the correlation function for vanishing time delay (Champeney, 1973). Terms of higher order in this expansion can of course be introduced, but they are seldom used. Note that, for functions that are not differentiable, we can still define a correlation function; this will still be symmetric, $R(t) = R(-t)$, but will most likely be characterized by a cusp ('infinite curvature') at the origin. Singularities in the derivatives of a signal can be represented by δ-functions, and, when we take the mean-square value of the signal, we expect to find that this average diverges, giving a divergent second term in (2.32).

2.4.3.1   Characteristic time scales

Usually, three different timescales are associated with the normalized auto-correlation function $R(\tau) \equiv \langle Y(t) Y(t + \tau) \rangle / \langle Y^2 \rangle$ for time-stationary processes. One is the *integral timescale*, $t_1 \equiv \int_0^\infty R(\tau) \, d\tau$. The notation *micro timescale* is often seen for $t_m \equiv \sqrt{\langle Y^2 \rangle / \langle (dY/dt)^2 \rangle}$, which can be obtained from the curvature of the normalized auto-correlation function at zero time delay, see (2.32). Finally, a *correlation time* can be defined as the time value t_C, where $R(t) \approx 0$ when $t > t_C$. It is readily realized that $t_C \geq t_1$, in general.

Evidently t_1 and t_m are well defined, but their physical interpretations are not entirely obvious. Although it is possible to construct cases for which $R(t) = 0$ for some $t > t_C$, we will usually find that $R(t) \neq 0$ for all t, although $R(t)$ generally becomes very small for large time separations t. The value of t_C is therefore not well defined in the general case. Its physical interpretation for ergodic Gaussian random signals is, on the other hand, quite explicit; two signal values obtained with time separations $t > t_C$ in a given record are statistically independent, see Section 2.4.5. It is often assumed (and often erroneously) that two values of a signal can be considered statistically independent when they are obtained at time separations larger than a value for which the correlation has become very small but not necessarily zero.

2.4.4   'Almost' stationary processes

The stationary random process is an important concept, but evidently it can never be realized in practice; there will always be variations in one or all of the many parameters which usually determine the actual signal. Fluctuations in electric circuits depend on the temperature, etc. and after all we have to turn the device on at some time, and it can not be operating for all later times either, so nonstationary conditions are inevitable in practice. Usually one introduces the idea of *locally* time-stationary processes, or slowly varying stationary processes, in spite of the concept being just as meaningless as that of a slightly bent straight line, strictly speaking. Here we shall attempt to give at least *some* meaning to the idea.

Consider first a nonstationary random process, $Y(t)$. By ensemble averaging, it is still possible to obtain a correlation function $R(t_1, t_2) \equiv \langle Y(t_1)Y(t_2)\rangle$, but this function will now depend explicitly on t_1 and t_2, rather than just on the separation $t_1 - t_2$. Introducing the average time $\Upsilon = \frac{1}{2}(t_1 + t_2)$ and separation time $\tau = t_1 - t_2$, the correlation function is rewritten in terms of these new variables as $R(\Upsilon, \tau)$, with $t_1 = \Upsilon + \frac{1}{2}\tau$ and $t_2 = \Upsilon - \frac{1}{2}\tau$. If $R(\Upsilon, \tau)$ varies much faster with τ than it does with Υ, we can argue that the process is locally time stationary, at least in the wide sense discussed before in this section.

To complete the discussion we also consider the correlation function for the derivative process, i.e.

$$R_D(t_1, t_2) \equiv \left\langle \frac{d}{dt_1} Y(t_1) \frac{d}{dt_2} Y(t_2) \right\rangle$$

$$= \frac{\partial^2}{\partial t_1 \partial t_2} \langle Y(t_1)Y(t_2)\rangle$$

$$\equiv \frac{\partial^2}{\partial t_1 \partial t_2} R(t_1, t_2). \tag{2.33}$$

Introducing again $R_D(\Upsilon, \tau)$ and using

$$\frac{d^2}{dt_1 dt_2} \Rightarrow \frac{1}{4}\frac{\partial^2}{\partial \Upsilon^2} - \frac{\partial^2}{\partial \tau^2},$$

we find that

$$R_D(\Upsilon, \tau) = \frac{1}{4}\frac{\partial^2}{\partial \Upsilon^2} R(\Upsilon, \tau) - \frac{\partial^2}{\partial \tau^2} R(\Upsilon, \tau), \tag{2.34}$$

with $R(\Upsilon, \tau)$ as given before.

Assuming now that we *do* have a stationary random process, we find in particular that

$$R_D(\Upsilon, \tau) = R_D(\tau) = -\frac{d^2}{d\tau^2} R(\tau), \tag{2.35}$$

which is a particularly simple relationship between the correlation function for a stationary random process $Y(t)$ and that for its time-derivative process, $dY(t)/dt$. The result (2.35) has evident generalizations for higher derivatives. For an almost-stationary process we expect the first term in (2.34) to be much smaller than the second. The nth derivative of $R(\tau)$ at $\tau = 0$ determines $\langle (dY(t)/dt)^{2n}\rangle$ in much the same way as that in which the nth derivative of the characteristic function at the origin determines $\langle Y^n(t)\rangle$.

2.4.5   Gaussian random processes

Gaussian random processes are particularly important because they adequately describe a large number of natural phenomena. For a brief outline see Appendix C. By a Gaussian random process we understand one with the property that, for every integer N and every set of points $Y(t_1)$, $Y(t_2)$, $Y(t_3)$, ..., $Y(t_N)$, its Nth-order probability density function has an N-dimensional normal form

$$P_N(y_1, t_1; \ldots; y_N, t_N) = (2\pi)^{-N/2}|A|^{-1/2}\exp\left(-\frac{1}{2|A|}\sum_{i,j=1}^{N}A_{ij}y_i y_j\right), \tag{2.36}$$

where A_{ij} is the cofactor of $R(t_i, t_j)$ in the correlation matrix

$$||A|| = \begin{Vmatrix} R(t_1, t_1) & R(t_1, t_2) & \cdots & R(t_1, t_N) \\ R(t_2, t_1) & R(t_2, t_2) & \cdots & R(t_2, t_N) \\ \vdots & \vdots & \vdots & \vdots \\ R(t_N, t_1) & R(t_N, t_2) & \cdots & R(t_N, t_N) \end{Vmatrix} \tag{2.37}$$

and $|A|$ is the determinant of the matrix. Evidently, we are here dealing with square matrices, i.e. ones with the same number of rows and columns. The cofactor A_{ij} of any element $R(t_i, t_j)$ is defined to be the determinant of order $N - 1$ formed by omitting the ith row and jth column of $||A||$ and then multiplying it by $(-1)^{i+j}$. Without loss of generality, we assumed from the outset that the signal has zero mean, $\langle Y(t) \rangle = 0$, since this can always be achieved by a proper change of variables for stationary processes, see also Appendix C.

Generally, a Gaussian random process is completely described by its average value and its auto-correlation function $R(t_i, t_j)$, since it completely specifies the joint probability density in this case. For stationary processes, as mentioned, $R(t_i, t_j) = R(|t_i - t_j|)$. The auto-correlation function is obtained from the joint probability density as

$$R(t_i, t_j) \equiv \langle Y(t_i)Y(t_j) \rangle = \int\int_{-\infty}^{\infty} y_i y_j P_2(y_i, t_i; y_j, t_j) \, dy_i \, dy_j, \tag{2.38}$$

with $y_i = y(t_i)$ and $y_j = y(t_j)$. In this case, P_2 is the bivariate Gaussian obtained from (2.36) with $N = 2$.

- **Example:** For illustration consider first $N = 1$,

$$P_1(y_1, t_1) = (2\pi)^{-1/2}|A|^{-1/2}\exp\left(-\frac{A_{11}}{2|A|}y_1^2\right), \tag{2.39}$$

where for this simple case, $||A|| = ||R(t_1, t_1)||$ and $|A| = R(t_1, t_1)$, while $A_{11} = 1$. A less trivial case is $N = 2$, where

$$P_2(y_1, t_1; y_2, t_2) =$$
$$\frac{1}{2\pi\sqrt{|A|}}\exp\left(-\frac{1}{2|A|}(A_{11}y_1^2 + A_{22}y_2^2 + A_{12}y_1y_2 + A_{21}y_2y_1)\right). \tag{2.40}$$

Now the correlation matrix becomes

$$||A|| = \begin{Vmatrix} R(t_1, t_1) & R(t_1, t_2) \\ R(t_2, t_1) & R(t_2, t_2) \end{Vmatrix},$$ (2.41)

and $|A| = R(t_1, t_1)R(t_2, t_2) - R(t_1, t_2)R(t_2, t_1)$ is the determinant of the matrix. We have $R(t_1, t_2) = R(t_2, t_1)$. Then $A_{12} = A_{21} = -R(t_1, t_2)$. For time-stationary processes $A_{11} = A_{22} = R(t_1, t_1) = R(t_2, t_2)$. The arguments are readily generalized to the joint probability density associated with any number, N, see also Appendix C.

Assume now that for some reason the signals obtained at different times are uncorrelated. In that case all $R(t_i, t_j) = 0$ for $i \neq j$ and the Gaussian joint probability density becomes

$$P_N(y_1, t_1; \ldots; y_N, t_N) = (2\pi)^{-N/2}|A|^{-1/2} \exp\left(-\frac{1}{2}\sum_{i=1}^{N}\frac{y_i^2}{R(t_i, t_i)}\right).$$ (2.42)

The correlation matrix becomes diagonal for this case:

$$||A|| = \begin{Vmatrix} R(t_1, t_1) & 0 & \cdots & 0 \\ 0 & R(t_2, t_2) & \cdots & 0 \\ \vdots & \vdots & \vdots & \vdots \\ 0 & 0 & \cdots & R(t_N, t_N) \end{Vmatrix}.$$ (2.43)

Then $|A| = \Pi_{i=1}^{N} R(t_i, t_i)$, and $A_{ij} = 0$ if $i \neq j$ and $A_{ii} = |A|/R(t_i, t_i)$ if $i = j$. Hence

$$P_N(y_1, t_1; \ldots; y_N, t_N) = P_1(y_1)P_1(y_2)\cdots P_1(y_N).$$

Thus, in this case, the signals y_i and y_j are independent for $i \neq j$, emphasizing an important property, namely that uncorrelated Gaussian variables are also statistically independent. Intuitively we expect that, for a given ensemble of time records, the signals at two times t_1 and t_2 tend to become uncorrelated as $|t_1 - t_2| \to \infty$, and, consequently, for Gaussian processes, the two events also tend to be independent.

The latter expressions appear rather simple since they were obtained for the case in which the variable Y refers to *one* particular quantity, i.e. the individual records of the ensemble are sampled at N different times. The arguments are readily generalized to the case in which N signals of different origins are analyzed. The symmetry relation for the cross-correlation matrix becomes

$$||A||(\tau) = ||A||^t(-\tau),$$

where the superscript ^{t} denotes the transposed matrix. In cases in which the signals constitute a joint Gaussian random process, we can again argue that lack of correlation is a sign of statistical independence of the signals.

- **Exercise:** Demonstrate that $\langle Y(t_1)Y(t_2)Y(t_3)\rangle = 0$ and

$$\langle Y(t_1)Y(t_2)Y(t_3)Y(t_4)\rangle = \langle Y(t_1)Y(t_2)\rangle\langle Y(t_3)Y(t_4)\rangle$$
$$+ \langle Y(t_2)Y(t_3)\rangle\langle Y(t_1)Y(t_4)\rangle + \langle Y(t_2)Y(t_4)\rangle\langle Y(t_1)Y(t_3)\rangle,$$ (2.44)

for a Gaussian random process $Y(t)$ with zero mean. Does this relation assume *stationary* random processes?

- **Exercise:** Demonstrate that

$$\langle vf(v) \rangle = \langle v^2 \rangle \left\langle \frac{df(v)}{dv} \right\rangle, \tag{2.45}$$

when v is a Gaussian variable with zero mean (Furutsu, 1963, Novikov, 1964). Generalize this result, the Furutsu–Novikov theorem, to a centered Gaussian vector variable $\mathbf{v} = \{v_i\}$, $i = 1, 2, \ldots, n$.

2.5   The Wiener–Khinchine theorem

The importance attributed to the auto-correlation function is due to the interpretation of its Fourier transform as a power spectral density. In fact, in many ways it might be argued that the wisest thing to do is simply to *define* the (auto) power spectral density of a stochastic process as the Fourier transform of the auto-correlation function. (Similarly, the *cross-spectral* density or *cross-power spectrum* of *two* processes is the Fourier transform of their cross-correlation function.) Such a definition, no matter how appropriate it might be, does not provide much physical insight. In the following the concept of the auto-power spectrum is discussed in some detail.

Consider an arbitrary real-valued function of time, ${}^k x(t)$, with $-\infty < t < \infty$, where ${}^k x(t)$ is a sample function from a (possibly nonstationary) random process labeled by ^{k}. Following Bendat (1958), we define

$$
{}^k x_T(t) = \begin{cases} {}^k x(t) & |t| \leq T \\ 0 & \text{otherwise.} \end{cases} \tag{2.46}
$$

Assume that the Fourier transform of ${}^k x_T(t)$ exists and is denoted by

$$A_T(f, {}^k x) = \int_{-\infty}^{\infty} {}^k x_T(t) e^{-i2\pi ft} \, dt = \int_{-T}^{T} {}^k x(t) e^{-i2\pi ft} \, dt.$$

Assume that the power spectral density function $S_{XX}(f)$ exists, where

$$S_{XX}(f) = \lim_{T \to \infty} S_{XX}(f, T), \tag{2.47}$$

with

$$S_{XX}(f, T) = \langle S_T(f, {}^k x) \rangle,$$

where

$$S_T(f, {}^k x) = \frac{|A_T(f, {}^k x)|^2}{T}.$$

Define

$$
{}^k J_T(\tau) = \frac{1}{2T} \int_{-\infty}^{\infty} {}^k x_T(t) {}^k x_T(t + \tau) \, dt. \tag{2.48}
$$

Then, from (2.46), it is clear that

$$
{}^k J_T(\tau) = \frac{1}{2T} \int_{-T}^{T} {}^k x(t) {}^k x(t + \tau) \, dt + \mathcal{O}(1/T) \quad \text{as} \quad T \to \infty, \tag{2.49}
$$

where the notation $\mathcal{O}(1/T)$ indicates an error term that approaches zero as $T \to \infty$.

Assuming all mathematical operations in the following to be legitimate, we obtain, starting with the Fourier transform of $^k J_T(\tau)$,

$$\int_{-\infty}^{\infty} {}^k J_T(\tau) e^{-i2\pi f \tau} d\tau = \int_{-\infty}^{\infty} e^{-i2\pi f \tau} \left(\frac{1}{2T} \int_{-\infty}^{\infty} {}^k x_T(t) {}^k x_T(t+\tau) dt \right) d\tau$$

$$= \frac{1}{2T} \int_{-\infty}^{\infty} {}^k x_T(t) e^{i2\pi f t} dt \int_{-\infty}^{\infty} {}^k x_T(t+\tau) e^{-i2\pi f(t+\tau)} d\tau$$

$$= \tfrac{1}{2} S_T(f, {}^k x), \tag{2.50}$$

where we have in one place multiplied by $e^{i2\pi f t} e^{-i2\pi f t} = 1$, and in another place recognized that

$$\int_{-\infty}^{\infty} {}^k x_T(t+\tau) e^{-i2\pi f(t+\tau)} d\tau = \int_{-\infty}^{\infty} {}^k x_T(t) e^{-i2\pi f t} dt = A_T(f, {}^k x).$$

Thus, for a single arbitrary time record $^k x(t)$, with k fixed, we have

$$S_T(f, {}^k x) = 2 \int_{-\infty}^{\infty} {}^k J_T(\tau) e^{-i2\pi f \tau} d\tau. \tag{2.51}$$

Taking an average over all records k in the ensemble yields

$$S_{XX}(f, T) = 2 \int_{-\infty}^{\infty} \langle {}^k J_T(\tau) \rangle e^{-i2\pi f \tau} d\tau. \tag{2.52}$$

From (2.49) we have

$$\langle {}^k J_T(\tau) \rangle = \frac{1}{2T} \int_{-T}^{T} \langle {}^k x(t) {}^k x(t+\tau) \rangle dt + \mathcal{O}(1/T)$$

$$= \frac{1}{2T} \int_{-T}^{T} \gamma_{XX}(t, t+\tau) dt + \mathcal{O}(1/T), \tag{2.53}$$

as $T \to \infty$, where we introduced

$$\gamma_{XX}(t, t+\tau) = \langle {}^k x(t) {}^k x(t+\tau) \rangle \tag{2.54}$$

to represent the general nonstationary form of the auto-correlation function for the random process $X(t)$ at the times t and $t+\tau$. Thus, for any random process, as T becomes large, we derive the result

$$S_{XX}(f) = \lim_{T \to \infty} 2 \int_{-\infty}^{\infty} e^{-i2\pi f \tau} \left(\frac{1}{2T} \int_{-T}^{T} \gamma_{XX}(t, t+\tau) dt \right) d\tau. \tag{2.55}$$

The amount of work involved in using this general formula is in practice quite significant.

If the random process is stationary, a considerable simplification occurs because, as discussed before,

$$\gamma_{XX}(t, t+\tau) = \gamma_{XX}(0, \tau) = R_{XX}(\tau). \tag{2.56}$$

Then

$$\frac{1}{2T} \int_{-T}^{T} \gamma_{XX}(t, t+\tau) dt = R_{XX}(\tau), \qquad \text{independent of } T. \tag{2.57}$$

Hence, in this special case (2.55) becomes

$$S_{XX}(f) = 2 \int_{-\infty}^{\infty} R_{XX}(\tau) e^{-i2\pi f \tau} \, d\tau, \tag{2.58}$$

revealing a simple relationship between the power spectral density function and the auto-correlation function of a stationary random process. Except for a factor of two, $S_{XX}(f)$ is the Fourier transform of $R_{XX}(\tau)$. Since $R_{XX}(\tau)$ is an even function of τ, the above is equivalent to

$$S_{XX}(f) = 4 \int_{0}^{\infty} R_{XX}(\tau) \cos(2\pi f \tau) \, d\tau. \tag{2.59}$$

Also, from (2.58), the inverse Fourier transform of $S_{XX}(f)$ is

$$R_{XX}(\tau) = \frac{1}{2} \int_{-\infty}^{\infty} S_{XX}(f) e^{i2\pi f \tau} \, df, \tag{2.60}$$

and, using the fact that $S_{XX}(f)$ is an even function of f,

$$R_{XX}(\tau) = \int_{0}^{\infty} S_{XX}(f) \cos(2\pi f \tau) \, df. \tag{2.61}$$

From (2.61) it is easily demonstrated that $R_{XX}(\tau) = R_{XX}(-\tau)$, $R_{XX}(0) \geq |R_{XX}(\tau)|$ for all τ, and $dR_{XX}(\tau)/d\tau|_{\tau=0} = 0$ for time-stationary random processes, with the possible exception of processes with discontinuous derivatives.

In terms of angular frequency ω, in radians per second, (2.59) and (2.61) become

$$G_{XX}(\omega) = \frac{2}{\pi} \int_{0}^{\infty} R_{XX}(\tau) \cos(\omega \tau) \, d\tau \tag{2.62}$$

$$R_{XX}(\tau) = \int_{0}^{\infty} G_{XX}(\omega) \cos(\omega \tau) \, d\omega. \tag{2.63}$$

The last two formulas are known as the *Wiener–Khinchine relations*, named after two mathematicians, one American and the other Russian, who first noted the correspondence. (Owing to the multiplicity of ways of transliterating Russian names into the Roman alphabet, other spellings of the latter name can be seen.) The definition of the Fourier transform and its corresponding inverse transform is not unique; a factor $1/(2\pi)$ can to some extent be placed where it is most convenient (Champeney, 1973). All that *can* be required is that the transform followed by an inverse transform must reproduce the original function. When we introduce the normalized version $D_{XX}(\omega) = G_{XX}(\omega)/\int_{0}^{\infty} G_{XX}(\omega) \, d\omega$, with $\int_{0}^{\infty} G_{XX}(\omega) \, d\omega = R_{XX}(0)$, the quantity $D_{XX}(\omega)$ is sometimes interpreted as the probability density of finding a particular frequency ω, although this interpretation may be too simplistic.

Instead of the power spectral density functions $G_{XX}(\omega)$ and $S_{XX}(f)$, defined before, some writers define halves of these particular values to be the 'power spectral density functions,' namely

$$W_{XX}(\omega) = \tfrac{1}{2} G_{XX}(\omega)$$
$$W_{XX}(f) = \tfrac{1}{2} S_{XX}(f).$$

This notation makes the new functions $W_{XX}(\omega)$ and $W_{XX}(f)$ the exact Fourier transforms of the auto-correlation function $R_{XX}(\tau)$, see (2.58). These slightly different definitions for the same quantity appearing in the literature should not be confused (Bendat, 1958). Many formulas remain unaffected by this change since the same constant multiplying factor may appear on both sides of the equation.

Equations (2.59) and (2.61) or, equivalently, (2.62) and (2.63) are well-known relations between power spectral density functions and auto-correlation functions for stationary random processes. These formulas do not involve any 'ergodic hypothesis' as such, although this may actually be present by virtue of the manner in which the stationary auto-correlation function $R_{XX}(\tau)$ is calculated. By first determining $R_{XX}(\tau)$ and then using (2.59), we have a valid technique for computing $S_{XX}(f)$. This method was actually the one most often used until the fast-Fourier-transform (FFT, see e.g. Brigham (1974)) method was developed.

- **Example:** It is important to emphasize that in general $G(\omega = 0)$ is *not* the power in the dc value of a signal. For instance a signal X that is a constant varying randomly over the realizations is time stationary and has an auto-correlation function $R(\tau) = \langle x^2 \rangle$, which is constant for all τ. From the Wiener–Khinchine relations we find the corresponding power spectrum $G(\omega) = \langle x^2 \rangle \delta(\omega)$. A dc value is thus associated with δ-functions in the power spectra. On the other hand, the value $G(\omega = 0)$ for processes with *zero mean* is related to the integral of the auto-correlation function, see (2.62), giving $G(0) = (2/\pi) \int_0^\infty R(\tau)\,d\tau$.

    More generally, for time-stationary ergodic signals, two signal samples taken at times with large separation are likely to be independent, implying that

    $$\langle X(t)X(t+\tau)\rangle \to \langle X(t)\rangle\langle X(t+\tau)\rangle = \langle X(t)\rangle^2$$

    at large time separations. In other words, the auto-correlation function approaches a constant value at large time separations for signals with nonzero averages, $R(\tau \to \pm\infty) = \langle X(t)\rangle^2$. A nonvanishing average will also here imply a δ-function in the power spectrum.

    By choosing a proper reference value one might in principle always ensure that the average value of a given signal vanishes. In practice, this might not necessarily be a smart thing to do, as the example in Chapter 9 shows.

- **Example:** For a stationary random process we found a simple relation between the auto-correlation function for the process and the one for its time-derivative process, see (2.35). Using the Wiener–Khinchine relations, we find that the power spectrum, $G(\omega)$, associated with the time series $Y(t)$ itself, and the spectrum, $G_D(\omega)$, for the series obtained from its time derivative $dY(t)/dt$, are related by $G_D(\omega) = \omega^2 G(\omega)$.

The auto-correlation function is non-negative definite, as has already been mentioned, see (2.22). We generalize the result (2.22) by letting the index refer to continuous variables and obtain the often used, but slightly restrictive, form

$$\int\int_{-\infty}^{\infty} a(t_1)\langle Y(t_1)Y(t_2)\rangle a(t_2)\,dt_1\,dt_2 \equiv \int\int_{-\infty}^{\infty} a(t_1)R(t_1, t_2)a(t_2)\,dt_1\,dt_2 \geq 0. \qquad (2.64)$$

Assuming now that we have stationary processes, $R(t_1, t_2) = R(t_1 - t_2) \equiv R(\tau)$, and taking the special case of $a(t_{1,2}) = \exp(-i2\pi ft_{1,2})/\sqrt{2T}$ for $|t_{1,2}| \leq T$ and $a(t_{1,2}) = 0$ otherwise, we find in the limit of $T \to \infty$ that $S_{XX}(f) \geq 0$. This is a rather important result; it would be hardly meaningful to have a power spectral density that could assume negative values! Actually it can be demonstrated that any non-negative definite correlation function can be written as $R(\tau) = \int_{-\infty}^{\infty} S(\omega) \exp(-i\omega\tau)\, d\omega$ with $S(\omega) \geq 0$, this is *Bochner's theorem*.

- **Example:** Note that for instance a 'box function' $f(\tau) = A$ for $|\tau| \leq T$ and $f(\tau) = 0$ otherwise can not be acceptable as a correlation function even if it assumes only positive values; its Fourier transform is $2A \sin(\omega T)/\omega$. A box function is not non-negative definite. Similarly $f(\tau) = A \cos(\tau/T)$ for $|\tau| \leq \pi T/2$ and $f(\tau) = 0$ otherwise is not non-negative definite either.

- **Exercise:** Express the spectral density of the product of two real Gaussian (or *normal*) processes $Z(t) = X(t)Y(t) - \langle X(t)Y(t)\rangle$ in terms of the individual spectral densities by the relation

$$G_{ZZ}(\omega) = \int_{-\infty}^{\infty} G_{XX}(\omega - \omega')G_{YY}(\omega')\, d\omega'$$
$$+ \int_{-\infty}^{\infty} G_{XY}(\omega - \omega')G_{YX}(\omega')\, d\omega' + \langle X\rangle^2 G_{XX}(\omega) + \langle Y\rangle^2 G_{YY}(\omega).$$

  Alternatively, express the previous result in terms of correlation functions:

$$R_{ZZ}(\tau) = R_{XX}(\tau)R_{YY}(\tau) + R_{XY}(\tau)R_{YX}(\tau)$$
$$+ \langle X\rangle^2 R_{XX}(\tau) + \langle Y\rangle^2 R_{YY}(\tau).$$

  Write out the explicit expression for the special case of $X = Y$. Compare the results with (2.44).

- **Exercise:** Find the spectral density for the process $Z(t) = X(t)\, dX(t)/dt$ if X is a Gaussian random process and $R_{XX}(\tau) = A \exp(-\alpha|\tau|)\, [\cos(\beta\tau) + (\alpha/\beta)\sin(\beta|\tau|)]$, with $\langle X\rangle = 0$.

2.5.1   The Michelson interferometer

Because of the importance of the Wiener–Khinchine theorem, it may be worthwhile to consider a more 'physical' derivation. For this purpose we take a simple illustration based on the Michelson interferometer shown in Fig. 2.3. This experiment involves a light source, a detector, two perfectly reflecting mirrors, and a semitransparent mirror, which is here assumed to be ideal, i.e., to reflect/transmit exactly 50% of the incoming light. The mirror splits the light beam (full and dashed lines) into two branches. The lengths of the two branches are the same when the movable (top) mirror is at its reference position. The displacement of the mirror from this position is d.

Assume first that the source emits light at only two frequencies, ω_1 and ω_2. At the observation point the electric field is

$$E(t, l) = E_1[\cos(\omega_1 t - k_1 l) + \cos(\omega_1(t + \tau) - k_1 l)]$$
$$+ E_2[\cos(\omega_2 t - k_2 l) + \cos(\omega_2(t + \tau) - k_2 l)]$$
$$= 2E_1 \cos(\omega_1 t - k_1 l + \tfrac{1}{2}\omega_1\tau) \cos(\tfrac{1}{2}\omega_1\tau)$$
$$+ 2E_2 \cos(\omega_2 t - k_2 l + \tfrac{1}{2}\omega_2\tau) \cos(\tfrac{1}{2}\omega_2\tau),$$

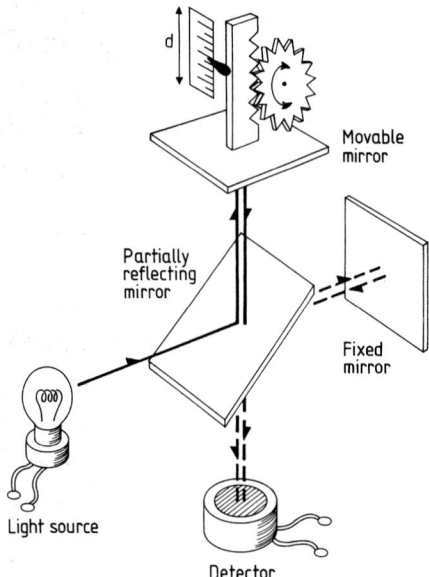

Figure 2.3 A schematic representation of a standard Michelson interferometer. A semitransparent, partially reflecting mirror splits the light beam (full and dashed lines) into two branches.

where $E_{1,2}$ is the field amplitude corresponding to $\omega_{1,2}$. Each field contains two contributions, originating from light being reflected by the movable (top) mirror and by the fixed mirror, respectively. The length of the reference branch is l while $\tau = 2d/c$ is the time delay between the two branches for light propagating at speed c, see Fig. 2.3. The energy density of the electric field at the observation point is $\frac{1}{2}\varepsilon_0 E^2(t, l)$. Similarly, we may obtain expressions for the magnetic field and calculate also the Poynting flux $E \times B/\mu_0$, etc. The coloring of a film placed at the observation position will be proportional to

$$
2T J_T = \int_{-T}^{T} E^2(t)\, dt
$$

$$
= 2T\left[2E_1^2 \cos^2\left(\frac{1}{2}\omega_1 \tau\right) + 2E_2^2 \cos^2\left(\frac{1}{2}\omega_1 \tau\right) \right]
$$

$$
+ 2E_1^2 \frac{1}{\omega_1} \cos^2\left(\frac{1}{2}\omega_1 \tau\right) \sin(2\omega_1 T) \cos(\omega_1 \tau)
$$

$$
+ 2E_2^2 \frac{1}{\omega_2} \cos^2\left(\frac{1}{2}\omega_2 \tau\right) \sin(2\omega_2 T) \cos(\omega_2 \tau)
$$

$$
+ 4E_1 E_2 T \cos\left(\frac{1}{2}\omega_1 \tau\right) \cos\left(\frac{1}{2}\omega_2 \tau\right)
$$

$$
\times \left[\frac{\sin[(\omega_1 - \omega_2)T]}{(\omega_1 - \omega_2)T} \cos\left((\omega_1 - \omega_2)\frac{\tau}{2}\right) \right.
$$

$$
\left. + \frac{\sin[(\omega_1 + \omega_2)T]}{(\omega_1 + \omega_2)T} \cos\left((\omega_1 + \omega_2)\frac{\tau}{2}\right) \right], \tag{2.65}
$$

where $2T$ is the opening time of the shutter. For simplicity and without loss of generality, we have here ignored the constant phase factor kl at the observation point. Evidently, the first term is proportional to T, while the second and third terms are $\mathcal{O}(1)$. If $(\omega_1 - \omega_2)T \gg 1$ then the fourth term will be of the same order. Assuming this to be the case we have for large T

$$J_T = (E_1^2 + E_2^2) + [E_1^2 \cos(\omega_1\tau) + E_2^2 \cos(\omega_2\tau)] + \mathcal{O}(1/T),$$

since $\cos^2(\frac{1}{2}x) = \frac{1}{2}(1 + \cos x)$. This result is readily generalized to the case of many (possibly infinitely many) discrete frequencies:

$$J_T = \sum_{j=0}^{N} E_j^2 + \sum_{j=0}^{N} E_j^2 \cos(\omega_j\tau) + \mathcal{O}(1/T), \tag{2.66}$$

still with the assumption that $(\omega_j - \omega_k)T \equiv T\Delta_{j,k}\omega \gg 1$ for *any* frequency pair (ω_j, ω_k). The accumulated intensity associated with one frequency, ω_j, is for large T proportional to $T E_j^2$, so the first term in (2.66) is just the sum of the spectral intensity contributions. The second term accounts for the interference (or 'beating') in the signal due to the time delay caused by the two paths of propagation.

The quantity J_T can now be calculated in a different way by noting that the electric field can be written as the sum of the fields of the directly transmitted wave and the wave delayed by the movable mirror, i.e. as

$$\begin{aligned} J_T &= \frac{1}{2T} \int_{-T}^{T} \frac{1}{2}[E(t) + E(t + \tau)]^2 dt \\ &\approx \frac{1}{2T} \int_{-T}^{T} E^2(t)\, dt + \frac{1}{2T} \int_{-T}^{T} E(t)E(t + \tau)\, d\tau, \end{aligned} \tag{2.67}$$

the latter approximation being valid for large T. Since the semitransparent mirror was assumed to be ideal, the two electric-field amplitudes are the same. The factor $\frac{1}{2}$ is introduced to have (2.67) agree with (2.65) at zero time delay, $\tau = 0$. The second term in (2.67) is the correlation function for time-stationary, ergodic processes. The two expressions for J_T, i.e. (2.66) and (2.67), must be identical, and the first term in (2.67) corresponds to the first term in (2.66). By equating the second term in (2.66) to the second term in (2.67), a discrete equivalent of the Wiener–Khinchine theorem is obtained. This physical explanation of the theorem may appear somewhat heuristic, but serves, among other things, to draw attention to the fact that an error is introduced by estimating a continuous power spectrum on the basis of a *finite* record length, since the requirement $\Delta_{j,k}\omega T \gg 1$ used when ignoring the last term in (2.65) can no longer be satisfied for all frequency separations.

- **Example:** Assume that the power spectrum consists of just one frequency. In a continuous representation the power spectrum is then $\frac{1}{2}E_0^2[\delta(\omega - \omega_0) + \delta(\omega + \omega_0)]$, whereas in the discrete form it is $E_j^2 = E_0^2$ for one particular value of j and $E_j^2 = 0$ otherwise. The two presentations differ in physical dimension for E_0^2, when we recall that $\delta(\omega)$ has the dimension of $1/\omega$, i.e. of time. The correlation function is $E_0^2 \cos(\omega_0\tau)$ for this example, irrespective of the interpretation of the power spectrum.

The Wiener–Khinchine theorem is explicitly derived for auto-correlation functions. Evidently, also cross-correlations can be Fourier transformed, but since the symmetry relation does not hold in this case, i.e. $R_{XY}(\tau) \neq R_{XY}(-\tau)$, the result is in general not a real function of frequency, and can not properly be interpreted as a power spectrum. Writing the Fourier transform of a cross-correlation function as $S_{XY}(\omega) = |A(\omega)| \exp(i\theta(\omega))$, one might interpret $|A(\omega)|$ as a cross-power spectrum and $\theta(\omega)$ as the cross-phase spectrum. We have $R_{XY}(\tau) = R_{YX}^*(-\tau)$ and $S_{XY}(\omega) = S_{YX}^*(\omega)$ by their construction.

- **Exercise:** Obtain the cross-power spectrum and cross-phase spectrum for the cross-correlation of the two signals $X(t)$ and $Y(t) = X(t + \Delta t)$. What is, in this case, the physical interpretation of these two quantities?

2.6   Asymptotic expansions

It is often an advantage to work with characteristic functions rather than probability distributions, in particular when studying analytic models for stochastic processes that are derived as sums of many independent processes. The probability density for the actual process is then obtained by taking the inverse Fourier transform of the resulting characteristic function. In reality this inverse transformation will usually not be possible in terms of simple analytic functions, and a closed expression for the probability density can not be readily obtained. In many cases, however, the detailed form of the probability density is not actually needed; often an asymptotic expansion of the probability density contains the desired information.

Similarly, we can often derive an expression for an auto-correlation function for a stationary random process for which the result is so complicated that it is difficult to obtain a closed analytic form for the power spectrum found by Fourier transformation of the correlation function. Again, we need not always be interested in the entire power spectrum; its asymptotic expansion will often suffice.

Asymptotic expansions of a function can be obtained from its Fourier transform. It may first be worthwhile to summarize some basic results. We write the Fourier transform as

$$\Psi(\omega) = \frac{1}{\sqrt{2\pi}} \int_{-\infty}^{\infty} \Phi(t) \exp(-i\omega t)\, dt$$

$$= \frac{1}{i\omega\sqrt{2\pi}} \int_{-\infty}^{\infty} \exp(-i\omega t) \frac{d}{dt} \Phi(t)\, dt$$

$$\vdots$$

$$= \frac{1}{(i\omega)^N \sqrt{2\pi}} \int_{-\infty}^{\infty} \exp(-i\omega t) \frac{d^N}{dt^N} \Phi(t)\, dt, \tag{2.68}$$

by use of partial integration, assuming that the differentials exist. Provided that the Fourier transform of $d^N \Phi(t)/dt^N$ exists, the last integral goes to zero as $1/\omega$ or faster, according to the Riemann–Lebesque theorem, independent of N. For a function that is infinitely many times differentiable, we therefore expect that its Fourier transform at large ω decays faster than $1/\omega^N$, for any N. Singularities or discontinuities in a function, or in its derivatives, which terminate the expansion (2.68) at a certain finite value of N are therefore particularly important for determining the asymptotic expansion of a Fourier transform.

A particularly simple discontinuity occurs when functions are multiplied by Heaviside's step function, $H(t)$. We can take a function $\Phi(t)$ and multiply it by $H(t - \alpha)$ and $H(\beta - t)$. Given, in general, an integral of the form $\Psi(\omega) = \int_\alpha^\beta \Phi(t) \exp(i\omega t)\, dt$, $\{\alpha; \beta\}$ being a real interval and $\Phi(t)$ a function that is continuously differentiable N times in this interval including the end points, it can be shown (Erdélyi, 1956) that, for $\omega \to \infty$,

$$\Psi(\omega) = B_N(\omega) - A_N(\omega) + \mathcal{O}(1/\omega^N), \tag{2.69}$$

where

$$A_N(\omega) = \sum_{n=0}^{N-1} i^{n-1} \frac{1}{\omega^{1+n}} \exp(i\omega\alpha) \frac{d^n}{dt^n} \Phi(t) \bigg|_{t=\alpha},$$

$$B_N(\omega) = \sum_{n=0}^{N-1} i^{n-1} \frac{1}{\omega^{1+n}} \exp(i\omega\beta) \frac{d^n}{dt^n} \Phi(t) \bigg|_{t=\beta}. \tag{2.70}$$

The corresponding result for $\omega \to -\infty$ is obtained by replacing t by $-t$. The proof of (2.69) is omitted here. For practical application, we can often use symmetry conditions, and the integration interval will typically be $\{\alpha; \beta\} = \{0; \infty\}$.

The theorem (2.69) can be generalized (Erdélyi, 1956, Lighthill, 1964) to

$$\int_\alpha^\beta (t-\alpha)^{\lambda-1}(t-\beta)^{\mu-1}\Phi(t)\exp(i\omega t)\,dt = B_N(\omega) - A_N(\omega) + \mathcal{O}(1/\omega^N), \tag{2.71}$$

with noninteger λ and μ in the interval $\{0; 1\}$, where

$$A_N(\omega) = \sum_{n=0}^{N-1} i^{n-1} \frac{\Gamma(n+\lambda)}{n!} \exp[i\omega\alpha + i\pi(n+\lambda-2)/2] \frac{1}{\omega^{n+\lambda}}$$

$$\times \frac{d^n}{dt^n}\left[(\beta-t)^{\mu-1}\Phi(t)\right]\bigg|_{t=\alpha},$$

$$B_N(\omega) = \sum_{n=0}^{N-1} i^{n-1} \frac{\Gamma(n+\mu)}{n!} \exp[i\omega\beta + i\pi(n+\mu-2)/2] \frac{1}{\omega^{n+\mu}}$$

$$\times \frac{d^n}{dt^n}\left[(t-\alpha)^{\lambda-1}\Phi(t)\right]\bigg|_{t=\beta}, \tag{2.72}$$

which actually includes the foregoing result for $\lambda = 1$ and $\mu = 1$. Generalizations of (2.71) can be obtained for asymptotic expansions of functions containing $|t|^\lambda$, $t^\lambda \log(|t|)$ and $|t|^\lambda \log(|t|)$ with λ an integer or noninteger, see Lighthill (1964) for results.

- **Exercise:** Determine the leading term in the asymptotic expansion of the Fourier transform of $\exp(-|t|)$ and compare it with the exact result which can be found in standard tables of Fourier transforms.

3 Fluctuations in electric circuits

It was demonstrated experimentally first by Johnson (1928) that electric networks in thermal equilibrium exhibit fluctuating potentials. An elegant theoretical analysis of the phenomenon was presented at the same time by Nyquist (1928) and the two works were published in sequence in a seminal issue of a scientific journal. In this chapter a simple physical example, directly related to the original problem posed by Johnson (1928), is discussed in detail. This particular model for an electric circuit is sufficiently simple to allow a transparent demonstration of the necessity and the consequences of fluctuations in thermal equilibrium. The essential constituents of the model are, apart from the assumption of thermal equilibrium, a simple, yet physically reasonable, statistical model for collisions between the atoms and molecules in the circuit.

3.1   A simple example

As an introduction to fluctuations in thermal equilibrium we consider a simple, physically realistic, example that can be analyzed in detail. A large capacitor made from two plates of area $\mathcal{A}$ with separation $\mathcal{L}$ is immersed into a background of an inert gas. Assume that one pair of charged particles, one electron and one ion, is placed between the plates. The charged particles are in thermal equilibrium with the gas and participate in the thermal motion. If we assume the two terminals of the capacitor to be short-circuited, as in Fig. 3.1, there will be no accumulation of charge on the plate and the electron and the ion will propagate along straight orbits between collisions with the neutral atoms. During such time intervals any of the charged particles will induce a current in the circuit by inducing charges q on the capacitor plates. This fluctuating current can in principle be recorded by standard methods and it can be analyzed in statistical terms.

The charge q induced by one particle can be determined by the following argument for the open circuit (without short-circuiting); the amount of work, W, done to induce a charge q on a plate, when the potential across the capacitor is V_c, must equal the energy U gained (or lost) by a charged particle moving through a certain potential difference V. For $W = qV_c$ we have $U = eV = eV_cz/\mathcal{L}$, i.e. $q = ez/\mathcal{L}$ with $e < 0$ being the electron charge. The current is obtained as $i = dq/dt$, giving the intuitively reasonable result

$$i = ew/\mathcal{L}, \tag{3.1}$$

w being the velocity component perpendicular to the plates, taken to be the z-direction. For the *open* circuit this current is exactly cancelled by the displacement current, i.e. the $\partial E/\partial t$ term in Maxwell's equations. When the capacitor plates are short-circuited, the current i is induced in the wire connecting the two plates. Note that the electron does not contribute particularly much

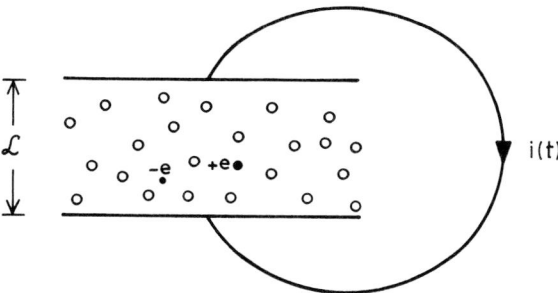

Figure 3.1 A simple physical model consisting of a pair of capacitor plates of area $\mathcal{A}$, separated by a distance $\mathcal{L}$. Neutral gas atoms are indicated by open circles.

to the current at the instant when it hits a capacitor plate; at that time it merely neutralizes the induced charge. An actual oscilloscope trace of the current as a function of time could look like the illustration in Fig. 3.2 with the assumption that the durations of the individual collisions are vanishingly small. The discontinuities arise when the electron collides with a neutral species, and the horizontal parts correspond to the free flights. The ion contribution to (3.1) will be ignored entirely in the following, since the characteristic velocity of an ion is much smaller than that of the electron, and the role of the ion will be to maintain overall charge neutrality.

Assume that a long record like that shown in Fig. 3.2 with a duration T is available. The distance between the capacitor plates is assumed to be so large that the possibility of the electron being absorbed by one of the plates within the time T can be ignored. The current is expanded in a Fourier series as

$$i(t) = I_0 + \mathrm{Re}\left(\sum_{\ell=1}^{\infty} \sqrt{2}\, I_\ell e^{i2\pi\ell t/T} \right), \tag{3.2}$$

where

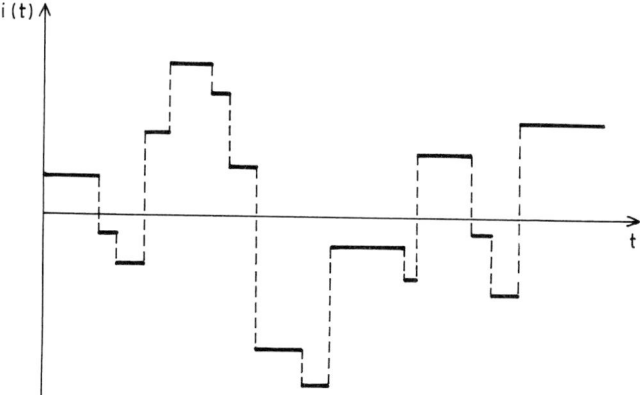

Figure 3.2 An idealized representation of the time variation of the current in the circuit in Fig. 3.1. Times at which the electron collides with neutral atoms are marked by vertical dashed lines.

$$I_\ell = \frac{\sqrt{2}}{T} \int_0^T i(t) e^{-i2\pi\ell t/T} \, dt,$$

$$I_0 = \frac{1}{T} \int_0^T i(t) \, dt. \tag{3.3}$$

The coefficient $\sqrt{2}$ is introduced to make I_ℓ the complex effective value of the current corresponding to the frequency $2\pi\ell/T$. This a standard convention. Assume that there are N collisions in the record. (The end points at $t = 0$ and $t = T$ need not correspond to collisions.) Then

$$I_0 = \frac{1}{T} \sum_{s=1}^N i_s$$

$$I_\ell = \frac{\sqrt{2}}{T} \left(\int_0^{t_1} i_1 e^{-i2\pi\ell t/T} \, dt + \int_{t_1}^{t_2} i_2 e^{-i2\pi\ell t/T} \, dt + \cdots + \int_{t_{N-1}}^T i_N e^{-i2\pi\ell t/T} \, dt \right)$$

$$= \frac{\sqrt{2}}{T} \sum_{s=1}^N i_s f_s e^{-i2\pi\ell t_s/T}, \tag{3.4}$$

where

$$f_s = \frac{1 - e^{-i2\pi\ell(t_{s-1}-t_s)/T}}{i2\pi\ell/T}. \tag{3.5}$$

Assume now that an ensemble of records like the one shown in Fig. 3.2 is available. The ensemble average of the Fourier coefficients can then be calculated, giving $\langle I_\ell \rangle = 0$ for all ℓ since the direction of the electron velocity, and thus the sign of i_s, is completely arbitrary. A more informative result can be obtained by considering

$$|I_\ell^2| = \frac{2}{T^2} \sum_{s=1}^N \sum_{r=1}^N i_s i_r f_s f_r^* e^{-i2\pi\ell(t_s-t_r)/T}, \tag{3.6}$$

where an asterisk indicates complex conjugation. Recall that i_r is a real quantity, $i_r^* = i_r$. The relation (3.6) contains several statistically varying quantities. The number of collisions N in a time record of given duration T thus varies from realization to realization. Similarly, all the times of the sth collisions, t_s, and the actual values of the currents, i_s, in the corresponding time interval vary also. Calculation of the ensemble average of $|I_\ell^2|$ requires knowledge of the statistical distributions of these quantities.

Assume first that N can be taken to be constant. Take, for instance, that from the (infinitely) many realizations we select the subset of all those having exactly N collisions. It is assumed that the z-component of the electron velocity, w, after a collision is statistically independent of the value it had *before* the collision. This statement can not be exact; a fast electron colliding with a slowly moving neutral atom will experience only a modest change in energy, i.e. $v^2 + u^2 + w^2$ changes only a little, thus indicating a constraint on the variation of w. The angle between the velocity vector and the z-axis can, however, safely be assumed to vary randomly at the collisions, and at least to some extent the statistical assumption regarding the w-component can be justified.

By assumption, the value of the current i_r in a time interval $t_r - t_{r+1}$ is independent of the number of collisions prior to the interval in question. The average value of (3.6) can then be decomposed as

$$\langle |I_\ell^2|\rangle = \frac{2}{T^2}\sum_{s=1}^{N}\sum_{r=1}^{N}\langle i_s i_r\rangle \langle f_s f_r^* e^{-i2\pi\ell(t_s-t_r)/T}\rangle. \tag{3.7}$$

The currents i_r and i_s in different time intervals were assumed to be statistically independent with $r \neq s$. Consequently $\langle i_r i_s\rangle = \langle i_s^2\rangle\delta_{r,s}$, where $\delta_{r,s}$ is the Kronecker delta, and (3.7) is reduced to

$$\langle |I_\ell^2|\rangle = \frac{2}{T^2}\sum_{s=0}^{N}\langle i_s^2\rangle \langle f_s f_s^*\rangle. \tag{3.8}$$

The central assumption of thermal equilibrium is introduced when calculating $\langle i_s^2\rangle$. The electron is participating in the thermal motion at a temperature T, and the probability of finding the z-component of its velocity in the range $w, w + dw$ is given by the Maxwell distribution, i.e.,

$$P(w)dw = \left(\frac{m}{2\pi\kappa T}\right)^{1/2} e^{-mw^2/(2\kappa T)}\, dw. \tag{3.9}$$

Here, $\kappa = 1.3805 \times 10^{-23}$ J K^{-1} is Boltzmann's constant. The transformation of variables $i = ew/\mathcal{L}$ gives

$$\langle i_s^2\rangle = \langle i^2\rangle = \int_{-\infty}^{\infty} i^2 \left(\frac{m}{2\pi\kappa T}\right)^{1/2}\frac{\mathcal{L}}{e}\exp\left(-\frac{m\mathcal{L}^2 i^2}{2e^2\kappa T}\right) di$$

$$= \frac{e^2}{m\mathcal{L}^2}\kappa T, \tag{3.10}$$

while $\langle i\rangle = 0$ as has already been stated. The system is time stationary in a statistical sense, meaning that no time t_s, with corresponding i_s, is from a statistical point of view distinguished from t_r, with i_r, when $r \neq s$. The remaining quantity to be calculated in the sum (3.8) is then $\langle |f_s|^2\rangle$, which depends solely on the collisional model.

3.2   The distribution of collision times

The collision between a charged particle and a neutral atom depends in general on their relative velocities in a complicated manner. In order to maintain the model on a tractable level it will here be assumed that the neutral species are so heavy that they might be considered almost immobile, at least compared with the moving free electron. Let the probability of a collision in a short time interval $\{t; t + \Delta t\}$ be

$$P(\text{one collision in } \Delta t) = v\,\Delta t, \tag{3.11}$$

v being a constant independent of the velocity of the electron. Effectively it is thus assumed that the collisional cross-section is inversely proportional to the relative velocity of the particle. Although this is a nontrivial restriction, it is not unphysical, and is a good approximation to a number of relevant processes. In effect it is assumed that the collisional cross-section is proportional to the interaction time, i.e. the time an electron spends in the vicinity of the neutral species, which is not an unreasonable assumption. These are not hard-sphere (or 'billiard-ball') collisions.

By assumption, the electron leaves the neutral species with a velocity component w entirely independent of the one with which it arrived. There is no persistence of velocities. The constant v is therefore independent of the 'history' of the electron, i.e. of foregoing collisions. This is a simple example of a Markov process, which is frequently encountered in practical applications. With the present model (3.11) the probability of a free flight starting at $t = 0$ and persisting at a time t can be calculated as follows (see also Appendix B).

Consider a time interval of length $t + \Delta t$ broken up into two subintervals, one of duration t and one of duration Δt. Since the probability of a collision in Δt is independent of the prehistory, it follows that $P(0, t + \Delta t) = P(0, t)P(0, \Delta t)$, where for brevity we introduced the notation $P(0, t + \Delta t) = P$ (no collision in the interval $t + \Delta t$), etc. Evidently $P(0, \Delta t) = 1 - v\Delta t$, the probability of two or more collisions in the time interval Δt being negligible when Δt is small. Consequently

$$\frac{P(0, t + \Delta t) - P(0, t)}{\Delta t} = -vP(0, t).$$

In the limit $\Delta t \to 0$ this difference equation becomes

$$\frac{dP(0, t)}{dt} = -vP(0, t), \tag{3.12}$$

giving the result

$$P(0, t) = e^{-vt}, \tag{3.13}$$

with the condition $P(0, t = 0) = 1$ from (3.11). This is the result for *no* collisions in the interval $\{0; t\}$. The probability of K collisions is given by the Poisson distribution

$$P(K, t) = \frac{(vt)^K e^{-vt}}{K!},$$

derived in Appendix B. The average number of collisions in a time interval T is obtained as

$$\langle K \rangle = \sum_{K=0}^{\infty} K \frac{(vT)^K e^{-vT}}{K!} = vT. \tag{3.14}$$

Evidently, this average need not be an integer number, even though K *is* an integer.

The average time between collisions can be found by noting that the probability of a free flight beginning at t_1 pertaining at $t_1 + t$ is $\exp(-vt)$. The probability that a free flight beginning at t_1 is terminated by a collision in the time interval $\{t_1 + t; t_1 + t + dt\}$ is then $\exp(-vt) v\, dt$. The average time between collisions is consequently given by

$$\int_0^{\infty} t \exp(-vt) v\, dt = \frac{1}{v}.$$

The average collision frequency is thus v, consistent with (3.14). The argument can be applied when 'looking backward' in time as well. Then $\exp(-vt) v\, dt$ is the probability that an ongoing free flight at t_1 started with a collision in a narrow time interval dt at $t_1 - t$.

3.2.1   An apparent paradox

When calculating the average time between collisions, it was implicitly argued that, because of the lack of memory of the process, one might as well take the time of the latest collision as the start of the observation period. Whatever the duration of the actual free flight, the residual time until the next collision was assumed to remain unaffected by the past and to have the same distribution as the free flight itself, giving the expected time to the collision as v^{-1}. Alternatively, it could, however, be argued that the probability of a free flight being terminated by a collision is the same for cases in which the observation is taken to begin at a time chosen at random between two consecutive collisions. For reasons of symmetry the expected or average time to the next collision should then be $\frac{1}{2}v^{-1}$. Both arguments seem perfectly reasonable and the apparent contradiction originates in the fact that the two discussions refer to somewhat different problems. In the second argument one particular element is chosen from the set of inter-arrival times $T_s = t_s - t_{s+1}$, namely the one such that $t_s > t > t_{s+1}$ for a selected time t, i.e. we are here dealing with a conditional probability. It turns out that the expectation of T_s is $2/v$ and the controversy is resolved. Heuristically, it can be argued that a long time interval has a better chance of covering t than does a short one (Zernike, 1929, Feller, 1971), see also Appendix B.

3.3   The mean-square Fourier coefficients

Using (3.5) and the arguments in Section 3.2, the ensemble average $\langle |f_s|^2 \rangle$ becomes

$$\langle |f_s|^2 \rangle = \int_0^T \frac{2[1 - \cos(2\pi\ell t/T)]}{4\pi^2\ell^2/T} ve^{-tv}dt$$

$$= \frac{2}{v^2 + (2\pi\ell/T)^2}, \tag{3.15}$$

where it was assumed that T is so large that the integration limits can be taken to infinity. The result (3.15) is independent of s, and the sum (3.8) is trivial, having N identical terms. The result is

$$\langle |I_\ell|^2 \rangle = 4\frac{N}{T^2}\kappa T\frac{e^2}{m\mathcal{L}^2}\frac{1}{v^2 + (2\pi\ell/T)^2}.$$

Hitherto, the number, N, of collisions in the time interval T was taken to be a constant. Quite generally, it is, however, also a statistically varying quantity, i.e. next time we consider a time interval of the same duration T we are likely to find a different number of collisions, N. The distribution of these numbers is known; it is the Poisson distribution which was discussed before. The average number of collisions in the time interval T is simply $\langle N \rangle = vT$. Averaging over all N, the foregoing expression is reduced to

$$\langle |I_\ell|^2 \rangle = 4\kappa T\frac{e^2}{m\mathcal{L}^2}\frac{v/T}{v^2 + \omega^2}, \tag{3.16}$$

where $\omega = 2\pi\ell/T$. Note that only positive frequencies are considered here since ℓ is confined to the interval $0 < \ell < \infty$. We could just as well use the interval $-\infty < \ell < \infty$ and use 2 instead of

4 in (3.16). Physically, the contribution from a negative frequency is indistinguishable from that from a positive one; we can not discriminate between clockwise, $e^{-i\omega t}$, and anticlockwise, $e^{i\omega t}$, rotation in the complex plane!

●   **Exercise:** Demonstrate that $\langle I_\ell I_s \rangle \to 0$ for $T \to \infty$ for $\ell \neq s$, i.e. that two different Fourier components are uncorrelated (but not necessarily statistically independent).

The mean-square values of the Fourier coefficients of the current are inversely proportional to the length of the time series. Simultaneously, the frequency resolution, i.e. the density of points on the frequency axis, is proportional to T. A result independent of T is obtained by considering the effective value of the current defined by $I_{\text{eff}}^2 = (1/T) \int_0^T i^2(t)\, dt$. By Parseval's theorem $I_{\text{eff}}^2 = \sum_\ell |I_\ell|^2$. The number of points on the ω-axis in an interval $\Delta\omega$ is $\Delta\omega T/(2\pi)$. The contribution to I_{eff}^2 from this narrow frequency band $\Delta\omega$ is then

$$I_{\text{eff}}^2(\omega; \omega + \Delta\omega) = \sum_{\ell = T\omega/(2\pi)}^{T(\omega + \Delta\omega)/(2\pi)} \langle |I_\ell|^2 \rangle$$

$$\approx \frac{2}{\pi} \kappa T \frac{e^2}{m\mathcal{L}^2} \frac{\nu}{\nu^2 + \omega^2} \Delta\omega, \tag{3.17}$$

obtained from (3.16), where it is here assumed that the frequency interval is so narrow that $\langle |I_\ell|^2 \rangle$ is essentially constant in $\Delta\omega$. In terms of frequency $f = \omega/(2\pi)$ the result (3.17) becomes

$$I_{\text{eff}}^2(f; f + \Delta f) = 4\kappa T \frac{e^2}{m\mathcal{L}^2} \frac{\nu}{\nu^2 + (2\pi f)^2} \Delta f. \tag{3.18}$$

The effective value of the fluctuating current in the system is thus expressed in terms of a collision frequency ν, which is in principle a measurable quantity. It will now be demonstrated that (3.18) can be expressed in terms of the frequency-dependent admittance of the system, which is even easier to obtain, being in principle a directly measurable macroscopic quantity.

3.4   The circuit impedance

The admittance of the empty capacitor is $i\omega C$. This is assumed to be unaffected by the inert gas, but is modified slightly by the presence of the one free electron. In order to evaluate this correction, we assume that an external voltage $V(t) = \text{Re}(Ve^{i\omega t})$ is applied across the capacitor terminals instead of the short-circuiting wire. Here, Re( ) denotes the real part of the expression in the parentheses. When this external voltage is applied, the system is no longer in thermal equilibrium!

The electric field between the two plates is now $E(t) = \text{Re}[(V/\mathcal{L})e^{i\omega t}]$. The electron's velocity is no longer constant between collisions but rather is given by

$$w(t) = w(t_1) - \text{Re}\left(\frac{eV}{im\mathcal{L}\omega} (e^{i\omega t} - e^{i\omega t_1}) \right),$$

where it is assumed that the electron starts out with a velocity $w(t_1)$ after a collision at $t = t_1$. The moving charge gives rise to a current through the generator

$$i(t) = \frac{ew(t_1)}{\mathcal{L}} - \text{Re}\left(\frac{e^2 V}{i\omega m\mathcal{L}^2}\left(1 - e^{-i\omega(t-t_1)}\right)e^{i\omega t}\right). \qquad (3.19)$$

This expression remains valid until the next collision. The first term accounts for the current which was discussed in the preceding subsection. The second term is also statistically varying with an average value different than zero. The statistical distribution of $\Delta t \equiv t - t_1$ is also given by $\exp(-v\,\Delta t)\,vd\,\Delta t$, since Δt denotes the time since the previous collision. All other quantities in the second term of (3.19) are deterministic and the average value of $i(t)$ is readily obtained as

$$\langle i(t) \rangle = \text{Re}\left(\frac{e^2 V}{m\mathcal{L}^2}\frac{e^{i\omega t}}{v + i\omega}\right), \qquad (3.20)$$

using $\int_0^\infty [1 - \exp(-i\omega\tau)]\exp(-v\tau)\,v\,d\tau = i\omega/(v + i\omega)$. The impedance $X(\omega)$ is obtained as the ratio of the externally applied (deterministic) voltage and the average value of the current response. The impedance is a macroscopic, nonfluctuating, quantity. The contribution from the free electron is

$$X(\omega) = v\frac{m\mathcal{L}^2}{e^2} + i\omega\frac{m\mathcal{L}^2}{e^2}.$$

Physically, the impedance $X(\omega)$ can be realized by putting a resistance $R = vm\mathcal{L}^2/e^2$ in series with an inductance $L = m\mathcal{L}^2/e^2$, i.e. $R = vL$. In addition comes the impedance $-i/(\omega C)$ of the empty capacitor, which is in parallel with this L–R series connection. (The impedance $-i/(\omega C)$ is the ratio of the externally applied voltage and Maxwell's displacement current $i\omega\varepsilon_0 AEe^{i\omega t} = i\omega\varepsilon_0 A(V/\mathcal{L})e^{i\omega t}$, recalling that $C = \varepsilon_0 A/\mathcal{L}$.)

The admittance of the entire system becomes

$$Y(\omega) = \frac{v}{v^2 + \omega^2}\frac{e^2}{m\mathcal{L}^2} + i\left(\omega C - \frac{\omega}{v^2 + \omega^2}\frac{e^2}{m\mathcal{L}^2}\right). \qquad (3.21)$$

On comparing this with (3.17) and (3.18) we obtain the rather remarkable result that

$$I_{\text{eff}}^2(\omega; \omega + \Delta\omega) = \frac{2}{\pi}\kappa T\,\text{Re}[Y(\omega)]\,\Delta\omega,$$

or, in terms of frequency, $f = \omega/(2\pi)$,

$$I_{\text{eff}}^2(f; f + \Delta f) = 4\kappa T\,\text{Re}[Y(f)]\,\Delta f, \qquad (3.22)$$

where ω and f are restricted to the interval $\{0, \infty\}$. This is a special case of the celebrated *Nyquist theorem*. It is remarkable because it relates two quantities of apparently very different origins: (i) a fluctuation in current with characteristics originating from a *microscopic* collisional process, and (ii) an admittance that characterizes a *macroscopic*, directly measurable, average quantity. Physically, the theorem states that the same processes, namely the collisions of the electron with neutral species, characterize the resistivity of the circuit and the fluctuations in current. The fluctuations are not *caused* by thermal fluctuations; they *are* thermal fluctuations. Now, this was a very special, carefully chosen, example. The theorem can be proven to be much more general, as will be demonstrated later on.

From the theorem for fluctuations in current just derived, we can obtain a corresponding result for fluctuations in voltage across the *open* circuit, namely that from which the short-circuiting wire in Fig. 3.1 is removed. The two expressions are related by Thévenin's and Norton's theorems (Horowitz and Hill, 1980). An equivalent circuit can be given (MacDonald, 1962) as the same ideal noise-free impedance as before, but now placed in *series* with an ideal voltage generator characterized by

$$V_{\mathrm{eff}}^2(\omega; \omega + \Delta\omega) = \frac{2\kappa T}{\pi} \, \mathrm{Re}\left(\frac{1}{Y(\omega)}\right) \Delta\omega \tag{3.23}$$

or

$$V_{\mathrm{eff}}^2(f; f + \Delta f) = 4\kappa T \, \mathrm{Re}\left(\frac{1}{Y(f)}\right) \Delta f. \tag{3.24}$$

Thévenin's and Norton's theorems simply state that any linear circuit element that includes one or more generators can be uniquely characterized by two measurable quantities; its short-circuit current and open-circuit voltage obtained as a function of frequency with attention to their phase relation. The ratio of the two quantities, which are, in general, complex, is the internal impedance of the element.

Before leaving this simple example it may be appropriate to point out that the generalization to the case in which there are *many* free charges in the system is quite straightforward, provided that the gas of charged particles is sufficiently dilute to allow the approximation that the charges are statistically independent, noninteracting charges. The total current is given by

$$i(t) = i_1(t) + i_2(t) + \cdots + i_N(t)$$

with, say, N electron–ion pairs between the capacitor plates. Since $\langle i_1(t)i_2(t)\rangle = \langle i_1(t)\rangle\langle i_2(t)\rangle = 0$ etc. for any pair of statistically independent electrons, we have

$$\langle i^2(t)\rangle = \langle i_1^2(t)\rangle + \langle i_2^2(t)\rangle + \cdots + \langle i_N^2(t)\rangle.$$

Evidently, all the electrons are in thermal equilibrium with the same system, so all standard deviations of their current contributions are identical, giving $\langle i^2(t)\rangle = N\langle i_1^2(t)\rangle$. Introducing the electron density $n = N/(\mathcal{A}\mathcal{L})$, we have

$$I_{\mathrm{eff}}^2(\omega; \omega + \Delta\omega) = \frac{2}{\pi}\kappa T \frac{\mathcal{A}}{\mathcal{L}} \frac{e^2 n}{m} \frac{\nu}{\nu^2 + \omega^2} \Delta\omega,$$

where $\mathcal{A}$ is the area of the plates. This result can be expressed as

$$I_{\mathrm{eff}}^2(\omega; \omega + \Delta\omega) = \frac{2}{\pi}\kappa T C \frac{\omega_{\mathrm{pe}}^2 \nu}{\nu^2 + \omega^2} \Delta\omega, \tag{3.25}$$

in terms of the capacitance $C = \varepsilon_0 \mathcal{A}/\mathcal{L}$ of the empty capacitor and the electron plasma frequency $\omega_{\mathrm{pe}} = [e^2 n/(\varepsilon_0 m)]^{1/2}$. By use of (3.23) we find

$$V_{\mathrm{eff}}^2(\omega; \omega + \Delta\omega) = \frac{2\kappa T}{\pi C} \frac{\nu\omega_{\mathrm{pe}}^2}{(\omega_{\mathrm{pe}}^2 - \omega^2)^2 + \nu^2\omega^2} \Delta\omega.$$

The capacitor, including gas and charge carriers, can evidently be considered as a generator of thermal noise with an internal, frequency-dependent, impedance. Norton's theorem (Horowitz and Hill, 1980) ensures that an equivalent circuit of the entire system can be written as the empty capacitor in parallel with a series connection of an ideal (noiseless) resistor of resistance $R = \nu/(C\omega_{\mathrm{pe}}^2)$ and an inductance $L = 1/(C\omega_{\mathrm{pe}}^2)$. The fluctuations can then be repre-

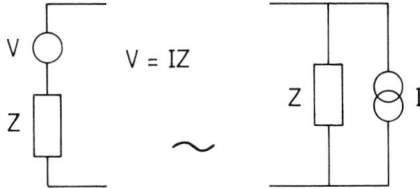

Figure 3.3 An illustration of Norton's and Thévenin's theorems. Thévenin's theorem states that any two-terminal network of resistors and voltage sources is equivalent to a *single* resistor in series with a single voltage source. This is remarkable; any mess of batteries and resistors can be mimicked with one battery and one resistor. When capacitors and inductors are present also, the theorem states that the two-terminal network is equivalent to a signal source without inner impedance in series with a complex impedance. Norton's theorem states that the same thing can be achieved with a current source in parallel with, in general, a complex impedance (Horowitz and Hill, 1980).

sented by an ideal current generator in parallel with the circuit, see Fig. 3.3. The characteristics of this current generator are known only in a statistical sense from the values of the mean-square Fourier coefficients of the current it imposes on the circuit.

It is interesting to note that, in the limit of small collision frequencies v, we have $v/(v^2 + \omega^2) \to \pi\delta(\omega)$ and the fluctuations in current are confined to low frequencies. For very large v, on the other hand, the fluctuations are distributed with constant power over a wide range of frequencies; $I_{\text{eff}}^2 \approx (2/\pi)\kappa TC(\omega_{\text{pe}}^2/v)\,\Delta\omega$ for $\omega \ll v$. For $v \to 0$ we have

$$V_{\text{eff}}^2 \to 2\frac{\kappa T}{\pi C}\left(\frac{\omega_{\text{pe}}}{\omega}\right)^2\delta\left[1 - \left(\frac{\omega_{\text{pe}}}{\omega}\right)^2\right] = 2\frac{\kappa T}{\pi C}\delta\left[1 - \left(\frac{\omega_{\text{pe}}}{\omega}\right)^2\right],$$

implying that in this limit the open circuit exhibits thermal fluctuations primarily around the plasma frequency.

3.4.1   Another look at the system impedance

It may be instructive to investigate the system impedance from a slightly different point of view than the one in the foregoing section. Consider again a medium composed of a large number of free electrons with density n, and assume also here for simplicity that the heavy ions are immobile. The frequency-dependent polarization of this medium can be obtained by calculating the average displacement of the electrons subject to an externally applied deterministic electric field $Ee^{i\omega t}$, which brings the system out of thermal equilibrium.

Under the influence of an external, deterministic, electric field, the average displacement, $\langle z \rangle$, of an electron is a deterministic quantity satisfying the differential equation

$$\frac{d^2\langle z \rangle}{dt^2} + v\frac{d\langle z \rangle}{dt} = \frac{e}{m}\,\text{Re}(E\,e^{i\omega t}). \tag{3.26}$$

The collisions give rise to an effective friction, which is characterized by the coefficient v. Note that v^{-1} is here a relaxation time that is introduced without reference to the actual underlying physical process. It is simply stated by equation (3.26) that the 'memory' of $\langle z \rangle$ decays with the time constant ν^{-1}. It can happen after one collision as in our physical model, but could just as well be due to many collisions giving partial contributions adding up to a decay time ν^{-1}.

The average displacement of the electron in the direction parallel to the electric field is obtained as

$$\langle z(t) \rangle = \frac{e}{m} \text{Re} \left(\frac{E}{i\omega} \frac{e^{i\omega t}}{\nu + i\omega} \right). \tag{3.27}$$

Introducing a complex notation, the polarization, P, of the medium is a deterministic quantity given by

$$P = en\langle z(t) \rangle = \frac{ne^2}{i\omega m} E \frac{e^{i\omega t}}{\nu + i\omega}, \tag{3.28}$$

giving the complex electric displacement $D = \varepsilon_0 E + P \equiv \varepsilon_0 \varepsilon E$, or $\varepsilon - 1 = P/(\varepsilon_0 E)$. For the present case the complex relative dielectric-permittivity function is consequently obtained as $\varepsilon(\omega) \equiv \varepsilon_1(\omega) + i\varepsilon_2(\omega) = 1 + i\omega_{pe}^2/[\omega(\nu + i\omega)]$. The real and imaginary parts of the relative permittivity are then

$$\varepsilon_1 = 1 - \frac{\omega_{pe}^2}{\nu^2 + \omega^2}, \tag{3.29}$$

$$\varepsilon_2 = -\frac{\nu\omega_{pe}^2}{\omega(\nu^2 + \omega^2)}, \tag{3.30}$$

with the plasma frequency $\omega_{pe} = [e^2 n/(\varepsilon_0 m)]^{1/2}$. Note that $\varepsilon_2 \simeq -1/\omega$ for $\omega \ll \nu$. Since electron–electron interactions are ignored, the foregoing results do not contain pressure effects, i.e. there are no consequences of a gradient in local electron density.

The voltage between the capacitor plates is $V = \mathcal{L} E e^{i\omega t}$ and the current through the capacitor is $A\, dD/dt$, which in this case becomes $i\omega A(\varepsilon_1 + i\varepsilon_2)\varepsilon_0 E e^{i\omega t}$. The admittance is the ratio of these two quantities, giving

$$Y(\omega) = -\frac{A}{\mathcal{L}} \varepsilon_0 \varepsilon_2 \omega + i\frac{A}{\mathcal{L}} \varepsilon_0 \varepsilon_1 \omega.$$

It is easily shown that this result agrees with the one obtained previously. $Y(\omega)$ is here the *response function* which relates the current in the system to an externally applied disturbance, here the applied voltage.

Note that the resistive component $\text{Re}[Y(\omega)]$ depends here solely on ε_2 and not at all on ε_1. The dissipation is here related to the imaginary part of the dielectric function. In thermal equilibrium we expect $\omega\varepsilon_2 < 0$ in order to have $\text{Re}[Y(\omega)] \leq 0$ at all frequencies in the range $\{-\infty, \infty\}$, corresponding to a dissipative circuit element, such as a resistor.

The average dissipated power per volume element is $\text{Re}(Vi)/(A\mathcal{L}) = \frac{1}{4}(Vi^* + V^*i) = -\frac{1}{2}\omega\varepsilon_0\varepsilon_2 |E|^2$; for a given time-varying applied electric field, the dissipation is solely determined by the dielectric losses, $\varepsilon_2 < 0$, caused by the collisional damping described by ν.

4   The fluctuation–dissipation theorem

In the foregoing chapter, the Nyquist theorem was illustrated by a specific example making it evident that, at least for that case, there is a simple relation between the dissipative properties of a circuit element and its associated thermal fluctuations in current. The relation is, however, quite general for systems in thermal equilibrium, as is demonstrated in the following. In its general form, the relation is called the fluctuation–dissipation theorem, for obvious reasons. In its general form, the theorem has been discussed for instance by Callen and Welton (1951) and Kubo (1957).

First a simple resonance circuit, shown in Fig. 4.1, is considered. It consists of a capacitance, C, in series with a self-inductance, L. The voltage across the capacitor is U and the charges on the plates are $\pm q$. The current flowing in the circuit is I. Evidently, $dq/dt = I$, $q = CU$, and $U = -L\, dI/dt$. The total energy in the system is $H = \frac{1}{2}CU^2 + \frac{1}{2}LI^2$. Since there are no dissipative elements, this energy is conserved. The dynamics of the circuit can (Symon, 1960) be described by the generalized coordinate q and the generalized momentum $p = LI$, since $dq/dt = \partial H/\partial p$ and $dp/dt = -\partial H/\partial q$. The equation of motion for the system becomes

$$\frac{d^2 q}{dt^2} + q\frac{1}{LC} = 0. \tag{4.1}$$

The Hamiltonian can be expressed as

$$H = p^2/(2L) + q^2/(2C). \tag{4.2}$$

For constant H, the trajectory of the system point is an ellipse in phase space, i.e. in the plane spanned by q and p.

Let the circuit be in thermal equilibrium with a heat reservoir at a temperature T. The probability of finding the system in a state within the small interval $\{p; p + dp, q; q + dq\}$ is then

$$P(q, p)\, dq\, dp = \frac{e^{-H(q,p)/(\kappa T)}}{\int \int_{-\infty}^{\infty} e^{-H(q,p)/(\kappa T)}\, dq\, dp}\, dq\, dp. \tag{4.3}$$

It is a basic statement of statistical mechanics (Jackson, 1968, Pathria, 1996, Reichl, 1998) that all microstates (q, p) are equally probable when they correspond to the same value of $H(q, p)$. Inserting (4.2), the normalization by the denominator of (4.3) is readily evaluated as $2\pi\kappa T\sqrt{LC}$. The probability density $P(q, p)$ can be expressed in terms of the variables U and I to obtain for instance

$$\langle U^2 \rangle = \int \int_{-\infty}^{\infty} U^2 P(U, I)\, dI\, dU = \frac{\kappa T}{C}, \tag{4.4}$$

$$\langle I^2 \rangle = \int \int_{-\infty}^{\infty} I^2 P(U, I)\, dI\, dU = \frac{\kappa T}{L}. \tag{4.5}$$

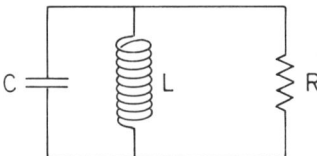

Figure 4.1 A simple resonance circuit consisting of a capacitance, C, and a self-inductance, L, placed in contact with a heat reservoir in the form of a large resistance, R, at temperature T. The thermal fluctuations which cause the excitations of the resonance circuit are formally taken into account by considering a current generator in parallel with an ideal resistance.

The average energy is $\langle H \rangle = \kappa T$. This is an example of the law of equipartition, stating that energy is distributed with $\frac{1}{2}\kappa T$ for each quadratic variable, here p and q.

These results are quite fundamental. We shall now discuss means of actually obtaining the thermal equilibrium, for which (4.4) and (4.5) can be applied. A heat reservoir is here realized by placing a large resistance, R, at temperature T in parallel with the resonant circuit in Fig. 4.1. The interaction must be weak, i.e. the resonance circuit must retain its 'identity'; the systems must be weakly coupled. This can be achieved if the resistor is sufficiently large, $R \gg \sqrt{L/C}$, so that the damping time is much larger than the resonance frequency, $1/\sqrt{LC}$. The trajectory of the system in phase space must be close to that of the ideal system for long times. The heat reservoir must have many more degrees of freedom than does the circuit. This is easily achieved; the resonator has only two. The thermal motion of the electrons in the resistor induces a fluctuating current through the resonance circuit. This effect can formally be taken into account by considering a current generator in parallel with an ideal noiseless resistor R. The statistical characteristics of this generator can then be determined by use of (4.4) and (4.5). Again the current is sampled in a long time interval, T, and represented by its Fourier series:

$$i(t) = I_0 + \mathrm{Re}\left(\sum_{\ell=1}^{\infty} \sqrt{2}\, I_\ell e^{i2\pi \ell t/T} \right). \tag{4.6}$$

Similarly, the fluctuating voltage across the circuit is given by

$$U(t) = U_0 + \mathrm{Re}\left(\sum_{\ell=1}^{\infty} \sqrt{2}\, U_\ell e^{i2\pi \ell t/T} \right). \tag{4.7}$$

The circuit is linear, so at any instant we have a simple relation between I_ℓ and U_ℓ provided by Ohm's law, giving

$$U_\ell = \frac{i\omega_\ell/\Omega}{1 - (\omega_\ell/\Omega)^2 + i(\omega_\ell/\Omega)d} Z_0 I_\ell, \tag{4.8}$$

where $Z_0 = \sqrt{L/C}$, $d = Z_0/R \equiv 1/Q$, $\Omega = 1/\sqrt{LC}$, and $\omega_\ell = 2\pi \ell/T$ are introduced for brevity. By Parseval's theorem we have $\langle U^2 \rangle = \sum_{\ell=1}^{\infty} \langle |U_\ell|^2 \rangle$, giving, by use of (4.4),

$$\frac{\kappa T}{C} = \sum_{\ell} \frac{(\omega_\ell/\Omega)^2}{[1 - (\omega_\ell/\Omega)^2]^2 + (\omega_\ell/\Omega)^2 d^2} Z_0^2 \langle |I_\ell|^2 \rangle. \tag{4.9}$$

The density of points on the frequency axis is very large when the time sequence is long. The number of points in a frequency interval $\Delta\omega$ is $\Delta\omega\, T/(2\pi)$. The sum in (4.9) can be approximated as

$$\frac{\kappa T}{C} = \int_0^\infty \frac{(\omega/\Omega)^2 Z_0^2 \langle\, |\, I_\ell\,|^2\rangle}{[1-(\omega/\Omega)^2]^2 + (\omega/\Omega)^2 d^2}\, \frac{T}{2\pi}\, d\omega. \tag{4.10}$$

Provided that d is small (Q is large), as we have initially assumed, we notice that the integrand is sharply peaked around the resonance frequency Ω, see also Fig. 4.2. If $\langle |I_\ell|^2\rangle$ varies slowly it can be placed outside the integral and we obtain, by introducing $x = \omega/\Omega$,

$$\begin{aligned}
\frac{\kappa T}{C} &= \frac{T}{2\pi} Z_0^2 \langle|I_\ell|^2\rangle\, \Omega \int_0^\infty \frac{x^2}{(1-x^2)^2 + x^2 d^2}\, dx \\
&= \frac{T}{2\pi} Z_0^2 \langle|I_\ell|^2\rangle \Omega \frac{\pi}{2d} \\
&= \frac{T}{4} \langle|I_\ell|^2\rangle \frac{R}{C},
\end{aligned} \tag{4.11}$$

giving

$$\langle|I_\ell|^2\rangle = 4\kappa T \frac{1}{R}\frac{1}{T}, \tag{4.12}$$

a result consistent with the assumption that $\langle|I_\ell|^2\rangle$ could be taken outside the integral; a simple resistor does not have any characteristic frequencies. By introducing the effective value of the current, a result similar to (3.22) can be obtained.

By reference to Norton's theorem, it can be concluded from (4.12) that the thermal fluctuations of the resistance R can be described by an equivalent diagram in which an ideal, noise-free resistance R is in parallel with a current generator supplying a fluctuating current such that $I_{\text{eff}}^2(f; f + df) = (4\kappa T/R)\, df$, where $df = 1/T$. From this result an expression for the fluctuations in open-circuit voltage can be obtained by application of Thévenin's theorem. The equivalent circuit can now be written (MacDonald, 1962) as the same resistor in series with a voltage generator characterized by

$$V_{\text{eff}}^2(f; f + df) = 4\kappa T R\, df. \tag{4.13}$$

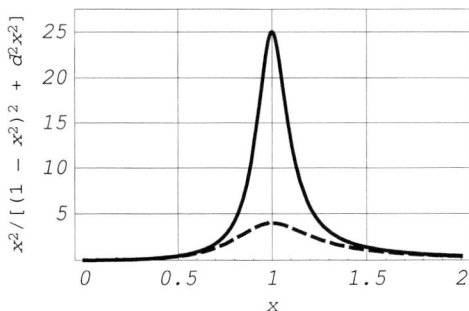

Figure 4.2 The weight function $x^2/[(1 - x^2)^2 + d^2 x^2]$, shown for $d = 0.2$ (full line) and $d = 0.5$ (dashed line). In terms of Dirac's δ-function, we have $\delta(x^2 - 1) = \lim_{d\to 0}(2d/\pi)x^2/[(1 - x^2)^2 + (xd)^2]$, see also Appendix D.

Thévenin's theorem is not trivial; it is valid also for a circuit element composed of a network of many passive (linear) components that, in thermal equilibrium, are associated with individual noise generators, i.e. a complicated electric network containing a variety of generators can be replaced by a single generator, whose voltage is the open-circuit output voltage of the system, in series with a single impedance, namely the internal impedance measured at the output terminals, with all the voltage generators short-circuited and current generators disconnected.

- **Example:** As a numerical example we take $R = 100$ kΩ, a bandwidth of 10 kHz and $T = 293$ K, i.e. room temperature, and find $V_{\text{eff}} \approx 4\,\mu$V, a rather small quantity.

Note that the fluctuation–dissipation theorem is *linear*; a doubling of R results in a doubling of V_{eff}^2. However, nonlinear circuit elements $R = R(V)$ can also be realized and such cases were excluded from the present derivation. Fluctuations in *nonlinear* circuit elements require a different treatment, whereby it is often necessary to consider each individual problem separately, see also Chapter 12.

It may be appropriate here to insert a word of caution. Some firms supply low-noise resistors, usually of metal-film type and fairly expensive. Now, Nyquist's theorem depends neither on the material nor on the manufacturer. Before suspecting a swindle, it is important to emphasize that the theorem depends strictly on the assumption of thermal equilibrium. This is no longer correct when a current is passed through a resistor, and the noise characteristics may change drastically. It is for instance important (and difficult, and hence expensive) to avoid small cracks in the material where small, noisy, 'sparks' can develop.

4.1   A *perpetuum mobile* of the second kind

The thermal fluctuation discussed in the foregoing tempts one to attempt a very simple construction of a *perpetuum mobile* of the second kind, i.e. a device that allows the transformation of heat to work involving only heat reservoirs at the same temperature. It could thus be argued that a small handy creature such as Maxwell's demon could extract energy from the thermal fluctuations to drive an electromotor, or charge a capacitor, by operating a switch when the polarity of the fluctuating voltage was favorable. This would imply that the actual voltage had to be detected and measured at all times, which can not be done without actually perturbing the system. However, the idea could seemingly be realized even by a passive element, a simple diode, which allows current to pass in one direction only. This scheme does not work either, since an ideal diode can not be realized in practice. In thermal equilibrium a physical diode is itself fluctuating and thus generating currents according to Nyquist's result. The current–voltage characteristic of a diode is

$$i = I_{\text{sat}}\left(e^{eV/(\kappa T)} - 1\right), \tag{4.14}$$

where I_{sat} is the saturation current. The conductivity is obtained as $di/dV = [eI_{\text{sat}}/(\kappa T)]\exp[eV/(\kappa T)]$ approximated as $di/dV \approx eI_{\text{sat}}/(\kappa T)$ for small V. The diode thus has an associated resistance and is consequently itself introducing a fluctuating current into the circuit, as described by the fluctuation–dissipation theorem.

4.2   The ultraviolet catastrophe

With $V_{\text{eff}}^2(f; f + \Delta f) = 4\kappa T R \, \Delta f$ an arbitrarily large value of V_{eff} can in principle be obtained for fixed Δf by choosing R sufficiently large. The available *power* in that bandwidth is, however, $\kappa T \, \Delta f$ independent of R since the internal resistor of the generator is itself given by R. With the assumption that an ideal resistor can be realized as constant for all frequencies, the spectrum is also constant, independent of frequency. This is, for obvious reasons, called *white noise*. It is readily noted, however, that, for this case, the integrated power $\int_0^\infty 4\kappa T \, df$ diverges; the high-frequency part of the spectrum contains infinite energy. This is called the *ultraviolet catastrophe*. It is not a practical problem really; an ideal resistor can not be realized for all frequencies. As $f \to \infty$, the self-induction of the connecting wires, capacitive coupling across terminals, etc. will become important. In particular, for the simple model example of Chapter 2, this divergence does not appear. However, logically this ultraviolet catastrophe was an uncomfortable puzzle and it turned out to be intimately connected to a basic shortcoming of classical physics. Its remedy was Planck's quantization.

The law of equipartition for the simple resonance circuit in Fig. 4.1 was derived in terms of the probability density $P(I, U)$. For this simple example $H = (LI^2 + CU^2)/2$ and the probability density can be easily expressed in terms of the Hamiltonian. The law of equipartition can then be obtained as

$$\langle H \rangle = \frac{\int_0^\infty H e^{-H/(\kappa T)} dH}{\int_0^\infty e^{-H/(\kappa T)} dH} = \kappa T.$$

The assumption implied is evidently that the energy of the system is continuously divisible. If it is now assumed that the energy is rather distributed to the oscillator only in multiples of an elementary quantum hf, where $h = 6.6256 \times 10^{-34}$ J s is Planck's constant, the average energy is obtained as a summation rather than an integral, i.e.

$$\langle H \rangle = \frac{\sum_{n=0}^\infty nhf e^{-nhf/(\kappa T)}}{\sum_{n=0}^\infty e^{-nhf/(\kappa T)}} = \frac{hf}{e^{hf/(\kappa T)} - 1}, \tag{4.15}$$

using $(1 - x)^{-1} = \sum_{n=0}^\infty x^n$ for $x < 1$. Nyquist consequently proposed a general expression for the effective value of the fluctuating voltage,

$$V_{\text{eff}}^2(f; f + df) = 4R \frac{hf}{e^{hf/(\kappa T)} - 1} \, df. \tag{4.16}$$

In the classical limit, obtained by formally letting $h \to 0$, this expression reduces to the one obtained previously. The inclusion of a zero-point energy $\frac{1}{2} hf$ can be argued, but this will be of little consequence for practical applications. A quantum-mechanical derivation of the fluctuation–dissipation theorem was given by Callen and Welton (1951).

- **Exercise:** Estimate the Planck frequency, $\kappa T / h$, at $T = 300$ K and $T = 3$ K. Will quantization of energy give rise to observable phenomena for audio amplifiers?

4.3   Dimensional arguments

The basic form of Nyquist's theorem could in principle be obtained by a dimensional argument. Thus the voltage V_{eff} is expressed in terms of the base MKSA units as $\text{m}^2\,\text{kg}\,\text{s}^{-3}\,\text{A}^{-1}$. The electric resistance is, on the other hand, $\text{m}^2\,\text{kg}\,\text{s}^{-3}\,\text{A}^{-2}$, while κT can be expressed in $\text{m}^2\,\text{kg}\,\text{s}^{-2}$. Now, any physical quantity can be written as the product of a constant Φ having the appropriate dimensions and a dimensionless function $F(\Theta)$ of a dimensionless variable Θ, see for instance Buckingham (1914) or Gibbings (1982). Thus, for instance, $V_{\text{eff}}^2/\Delta f = \Phi F(\Theta)$. With the convention that $[\Phi]$ denotes the dimension of Φ, this implies that

$$[\Phi] = (\text{m}^2\,\text{kg}\,\text{s}^{-3}\,\text{A}^{-1})^2(\text{s}).$$

Within a classical description V_{eff} can depend only on R, κT, and possibly frequency. We expect that Φ can be constructed from a combination of R, κT, and f, i.e.

$$[\Phi] = [R]^\alpha [\kappa T]^\beta [f]^\gamma$$
$$= (\text{m}^2\,\text{kg}\,\text{s}^{-3}\,\text{A}^{-2})^\alpha(\text{m}^2\,\text{kg}\,\text{s}^{-2})^\beta(\text{s}^{-1})^\gamma,$$

where the constants α, β, and γ are to be chosen such as to make the dimensions of the expressions correct, implying that $\alpha + \beta = 2$, $3\alpha + 2\beta + \gamma = 5$, and $\alpha = 1$. It is readily seen that the only possible choice is $\alpha = \beta = 1$ and $\gamma = 0$. A dimensionless quantity Θ can not be obtained by combining R, κT, and f, and hence $F(\Theta)$ must be a constant. Consequently, $V_{\text{eff}}^2/\Delta f \sim R\kappa T$, in agreement with Nyquist's result. The value 4 of the constant in the expression can of course not be determined by dimensional arguments.

If, however, quantization of energy is taken into account, Planck's constant enters the problem and now it becomes possible to construct a dimensionless variable $\Theta = hf/(\kappa T)$. By requiring that the classical result be obtained in the formal limit of $h \to 0$, we obtain the requirement that $\lim_{\Theta \to 0} F(\Theta) = \text{constant}$, in agreement with (4.16).

4.4   Frequency-dependent dissipative elements

The foregoing derivation of the fluctuation–dissipation theorem assumed that the resistance R which took the role of the heat reservoir was constant, independent of frequency. The resistance can, however, be the frequency-dependent $R(\omega)$ without consequence for the analysis, provided that the variation in ω is so slow that d can be considered constant in the integral in (3.11), i.e. $R(\omega)$ varies little in an interval of width d around Ω. In this case $R = R(\Omega)$ and $\langle |I_\ell|^2\rangle$ will similarly correspond to $\omega_\ell = \Omega$ in (3.12). In principle it can be argued that the entire noise characteristics of a given frequency-dependent resistor can be mapped out by a suitable collection of resonance circuits representing many different resonance frequencies. Still it must be required that $R(\Omega = 1/\sqrt{LC}) \gg Z_0 \equiv \sqrt{L/C}$ in all cases. For a given $R(\Omega)$ this can always be obtained by a proper choice of L and C. The result (3.12) is thus reproducing Nyquist's result from thermodynamic arguments without reference to the specific model of Chapter 3.

A frequency-dependent resistive element can be realized in many ways; Chapter 3 provided a particular, simple, example. Quite generally such an element can be realized by filling the volume between two condenser plates with a temporally dispersive medium, i.e. one for which the dielectric function is frequency dependent. The admittance for this circuit element was obtained as $Y(\omega) = -\varepsilon_0(\mathcal{A}/\mathcal{L})\omega[\varepsilon_2(\omega) - i\varepsilon_1(\omega)]$, giving

$$I_{\text{eff}}^2(\omega; \omega + d\omega) = -\frac{2}{\pi}\kappa T\varepsilon_0\frac{\mathcal{A}}{\mathcal{L}}\omega\,\varepsilon_2(\omega)\,d\omega, \tag{4.17}$$

and similarly

$$V_{\text{eff}}^2(\omega; \omega + d\omega) = -\frac{2}{\pi}\kappa T\frac{\mathcal{L}}{\varepsilon_0\omega\mathcal{A}}\frac{\varepsilon_2(\omega)}{\varepsilon_1^2(\omega) + \varepsilon_2^2(\omega)}\,d\omega, \tag{4.18}$$

with $0 \leq \omega < \infty$. For $-\infty \leq \omega < \infty$ we replace the multiplier $2/\pi$ by $1/\pi$. The coefficient $\mathcal{A}/\mathcal{L}$ refers to the actual physical arrangement, while the remaining parts of the expressions are solely associated with the material in question and its temperature. Evidently, the right-hand sides of (4.17) and (4.18) have to be positive, implying that $\varepsilon_2 \leq 0$. If the frequency range is extended to the range $\{-\infty; \infty\}$ we would require $\omega\varepsilon_2 \leq 0$ everywhere, i.e. $\varepsilon_2(-\omega) = -\varepsilon_2(\omega)$, to keep the power of the fluctuations positive.

The expression (4.18) can be put into a more convenient form by introducing the effective value of the electric field between the capacitor plates, $E_{\text{eff}} = V_{\text{eff}}/\mathcal{L}$:

$$\frac{1}{2}\varepsilon_0 E_{\text{eff}}^2(\omega; \omega + d\omega) = -\frac{\kappa T}{\pi\omega}\frac{1}{\mathcal{AL}}\frac{\varepsilon_2(\omega)}{\varepsilon_1^2(\omega) + \varepsilon_2^2(\omega)}\,d\omega, \tag{4.19}$$

where $\mathcal{AL}$ is the volume contained between the capacitor plates and ω is restricted to the interval $\{0; \infty\}$. The left-hand side of the expression (4.19) has the form of the energy density of an electrostatic field in vacuum. It might be somewhat misleading to introduce it in this context, without the modifications due to the presence of a dielectric (Landau and Lifshitz, 1960). Nevertheless, the formulation (4.19) is frequently used in the literature. An equivalent form has the volume $\mathcal{AL}$ as a multiplier on the left-hand side.

In active, or amplifying, media we can have $\varepsilon_2 > 0$, and oscillations can grow in amplitude, at least as long as the linearized description applies. Such conditions are, however, *not* stationary, and violate a basic assumption in the derivation of the fluctuation–dissipation theorem.

4.4.1   The dissipationless limit

Assume that we have control of a parameter that allows the transition $\varepsilon_2(\omega) \to 0$, at least in a certain frequency range. Consequently $I_{\text{eff}}^2 \to 0$ with $\varepsilon_2 \to 0$ according to (4.17) while

$$V_{\text{eff}}^2(\omega; \omega + d\omega) = 2\kappa T\frac{\mathcal{L}}{\varepsilon_0\mathcal{A}}\frac{1}{\omega}\delta(\varepsilon_1(\omega))\,d\omega, \tag{4.20}$$

using $\alpha/(\alpha^2 + x^2) \to \pi\delta(x)$ for $\alpha \to 0$ with $\alpha > 0$, see Appendix D.

As a specific example, consider again the results from Section 3.4.1, in which $\varepsilon_1(\omega) = 1 - \omega_{\text{pe}}^2/(v^2 + \omega^2)$ and $\varepsilon_2(\omega) = -v\omega_{\text{pe}}^2/[\omega(v^2 + \omega^2)]$, with v being a suitable control parameter for this case. One might vary the density of neutral gas between the condenser plates, for instance, and thus vary the collision frequency. For this particular example we thus find in the limit of $v \to 0$

$$I_{\text{eff}}^2(\omega; \omega + d\omega) = 2\kappa T \varepsilon_0 \frac{\mathcal{A}}{\mathcal{L}} \omega_{\text{pe}}^2 \delta'(\omega) \, d\omega, \tag{4.21}$$

using $\delta(x) = -x\delta'(x)$, see Appendix D. We find also

$$V_{\text{eff}}^2(\omega; \omega + d\omega) = 2\kappa T \frac{\mathcal{L}}{\varepsilon_0 \mathcal{A}} \frac{1}{\omega} \delta\left[1 - \left(\frac{\omega_{\text{pe}}}{\omega}\right)^2\right] d\omega,$$

or

$$V_{\text{eff}}^2(\omega; \omega + d\omega) = 2\kappa T \frac{\mathcal{L}}{\varepsilon_0 \mathcal{A}} \omega_{\text{pe}} \delta(\omega^2 - \omega_{\text{pe}}^2) \, d\omega. \tag{4.22}$$

We recognize $1 - (\omega_{\text{pe}}/\omega)^2$ as the simplest version of the plasma dielectric function for colli-sionless plasmas, see also (3.29).

The results (4.20) and (4.22) are somewhat uncomfortable by virtue of their demonstrating that the fluctuation–dissipation theorem predicts fluctuations in voltage at $\omega = \omega_{\text{pe}}$, or fluctua-tions in current at vanishing frequencies, $\omega = 0$, even in the absence of dissipation. It could first of all be argued with reference to the model of Chapter 3 that the limit $v \to 0$ is unphysical, since the electron will be rapidly absorbed by one of the condenser plates. However, it is at least formally quite feasible to increase $\mathcal{A}$ and $\mathcal{L}$ each time v is decreased, so the paradox can be formulated as the observation that (4.18) gives physically acceptable results for all v, no matter how small, except $v = 0$. It is, however, important to note that even this limit corresponds to a dissipative system in a certain sense. The fluctuations given by (4.22) are restricted to the plasma's resonance frequency ω_{pe}. If an oscillator is connected to the circuit with its frequency tuned exactly to ω_{pe}, it will supply energy to the system even when $v = 0$. This is true for any system with a resonance, be it a pendulum or an L–C circuit, driven at its resonance frequency. A general model equation is

$$\frac{d^2\eta}{dt^2} + \omega_0^2 \eta = \text{Re}(\varrho_0 e^{i\omega t}), \tag{4.23}$$

where ω_0 is a natural frequency of oscillation. At resonance, $\omega = \omega_0$, the dominant part of the solution at large times is

$$\eta = -\frac{t}{2\omega_0} \text{Re}(i\varrho_0 e^{i\omega_0 t}).$$

The linear increase of the amplitude with time is called *secular growth*. For instance, in the case in which (4.23) represents an L–C circuit with resonance frequency $\omega_0 = 1/\sqrt{LC}$, the electric-field energy in the condenser and the magnetic-field energy in the self-inductance increase linearly with time. This energy must evidently be supplied by the external generator, so the circuit will appear dissipative at this particular frequency, ω_0, in spite of the absence of dis-sipative elements such as resistors. It is interesting that the fluctuation–dissipation theorem treats this extreme limit consistently. The problem is important for discussions of thermal fluctuations in hot dilute plasmas. The basic equation is, for this case, the fully time-reversible Vlasov equation for the velocity distribution of particles. The collisionless damping for plasma oscillations, Landau damping, is thus of a different nature than those usually encountered, but it can be included in the analysis without constraints on the interpretation. It is, at least formally, possible to devise an equivalent circuit consisting of (infinitely many) capacitors and self-inductances, but no resistors, in a way that can mimic the collisionless dissipation of a plasma-filled capacitor (Pécseli, 1976).

4.4.2   Spatial and temporal dispersion

The fluctuating current and voltage, with effective values given before, are evidently related to the specific circuit element considered. It is in general preferable to have expressions that are related to the dielectric properties of the actual medium in the capacitor. If it can be assumed that the current density and electric fields are constant in the direction perpendicular to the condenser plates, this expression is readily obtained, as has already been demonstrated. Materials will, however, generally exhibit spatial as well as temporal dispersion. As a consequence, the fluctuating electric field will vary with time as well as with position. The dielectric function depends then on frequency and wavenumber as well, $\varepsilon(k, \omega) = \varepsilon_1(k, \omega) + i\varepsilon_2(k, \omega)$. On Fourier transforming the electric field in a (large) volume $A\mathcal{L}$ contained between the capacitor plates, see Fig. 3.1, we note that the density of components in wavenumber space is $A\mathcal{L}/(2\pi)^3$, and consequently the number of modes in a volume element in wavenumber space is $dk_x\, dk_y\, dk_z\, A\mathcal{L}/(2\pi)^3$. The rather self-evident generalizations of (4.17) and (4.19) are

$$J_{\text{eff}}^2(k; k + dk, \omega; \omega + d\omega) = -\frac{2}{\pi}\kappa T\omega\varepsilon_0\, \varepsilon_2(k, \omega)\, \frac{1}{(2\pi)^3}\, dk\, d\omega \qquad (4.24)$$

for the current density, and

$$\frac{1}{2}\varepsilon_0 E_{\text{eff}}^2(k; k + dk, \omega; \omega + d\omega) = -\frac{\kappa T}{\pi}\frac{1}{\omega}\frac{\varepsilon_2(k, \omega)}{\varepsilon_1^2(k, \omega) + \varepsilon_2^2(k, \omega)}\frac{1}{(2\pi)^3}\, dk\, d\omega \qquad (4.25)$$

for the electric field associated with wave-vectors and frequencies in a narrow interval $(dk, d\omega)$ around (k, ω) for fluctuations in thermal equilibrium, using the notation $dk = dk_x\, dk_y\, dk_z$.

The frequency dependence of the fluctuations is obtained by integrating over all wavenumbers, k. Apart from the trivial constant $A/\mathcal{L}$ the result corresponding to (4.18) is obtained for the case in which $\varepsilon(k, \omega) = \varepsilon(\omega)\delta(k)$. Note that (4.25) does not assume a priori the existence of a dispersion relation $\omega = \omega(k)$.

4.4.2.1   Anisotropic media

The foregoing discussion implicitly assumed that the dielectric media were isotropic, implying that the directions of, for instance, $E(k, \omega)$ and $D(k, \omega)$ are the same, and that $\varepsilon_0\varepsilon(k, \omega)$ is a scalar proportionality factor. In a more general case, for anisotropic media, we have $D(k, \omega) = \varepsilon_0\epsilon(k, \omega) \cdot E(k, \omega)$ with $\epsilon(k, \omega)$ being here a tensor with components $\epsilon_{j\ell}(k, \omega)$. For this case, for instance, (4.24) is generalized to (Bekefi, 1966)

$$J_j J_\ell^* \big|_{\text{eff}}(k; k + dk, \omega; \omega + d\omega) =$$
$$-\frac{2}{\pi}\kappa T\omega\varepsilon_0\frac{i}{2}\left[\epsilon_{\ell j}^*(k, \omega) - \epsilon_{j\ell}(k, \omega)\right]\frac{1}{(2\pi)^3}\, dk\, d\omega,$$

J_j being a component of the current-density vector J.

4.4.3   Longitudinal and transverse waves

The generalizations (4.24) and (4.25) seem rather self-evident, but deserve some scrutiny. A dielectric function associated with a medium reflects the way electrons and ions respond to electric fields. On the other hand, as we have seen, these very same electrons and ions generate electric fields during their random thermal motion. The synthesis of these two contributions is obtained by writing out Maxwell's equations including a fluctuation term:

$$\nabla \times H(r, t) = \tilde{J}(r, t) + \frac{\partial D(r, t)}{\partial t}, \tag{4.26}$$

$$\nabla \times E(r, t) = -\frac{\partial B(r, t)}{\partial t}, \tag{4.27}$$

$$\nabla \cdot D(r, t) = \tilde{\rho}(r, t), \tag{4.28}$$

$$\nabla \cdot B(r, t) = 0, \tag{4.29}$$

where $\tilde{\rho}$ is the fluctuating charge density and $\tilde{J}$ is the corresponding current density, where

$$\frac{\partial \tilde{\rho}(r, t)}{\partial t} + \nabla \cdot \tilde{J}(r, t) = 0. \tag{4.30}$$

The dielectric function or the dielectric susceptibility relating electric fields, E, and electric displacements, D, is obtained from the polarization charges by the standard analysis, see, e.g., Bekefi (1966) or Kittel (1968).

In a Fourier representation the equations (4.26)–(4.29) can be written as

$$ik \times H(k, \omega) = \tilde{J}(k, \omega) - i\omega D(k, \omega), \tag{4.31}$$

$$ik \times E(k, \omega) = i\omega B(k, \omega), \tag{4.32}$$

$$ik \cdot D(k, \omega) = \tilde{\rho}(k, \omega), \tag{4.33}$$

$$ik \cdot B(k, \omega) = 0, \tag{4.34}$$

with

$$i\omega \tilde{\rho}(k, \omega) - ik \cdot \tilde{J}(k, \omega) = 0. \tag{4.35}$$

By introducing the complex relative dielectric-permittivity function $\varepsilon(k, \omega) = \varepsilon_1(k, \omega) + i\varepsilon_2(k, \omega)$ and using (4.35), these relations can be rewritten as

$$k \times H(k, \omega) = i\tilde{J}(k, \omega) - \omega\varepsilon_0\varepsilon(k, \omega)E(\omega, k), \tag{4.36}$$

$$k \times E(k, \omega) = \omega B(k, \omega), \tag{4.37}$$

$$k \cdot E(k, \omega)\varepsilon(k, \omega) = i\frac{1}{\omega\varepsilon_0}k \cdot \tilde{J}(k, \omega), \tag{4.38}$$

$$k \cdot B(k, \omega) = 0. \tag{4.39}$$

The decomposition into a deterministic dielectric function and fluctuating charges and currents is formally equivalent to the presentation in Fig. 3.3, in which the fluctuating circuit element was represented by a passive element connected to a generator accounting for the fluctuations.

The fluctuation–dissipation theorem in its form (4.24) relates the fluctuations in current density to the dielectric losses. The current fluctuations can then be considered as 'driving' the fluctuations in the electric fields. From (4.36)–(4.39) the relation between the current density and the electric field is subsequently determined and the expression for the electric-field fluctuations is obtained. Two fundamentally different cases can be distinguished, namely electrostatic or longitudinal fluctuations and transverse electromagnetic fluctuations.

- **Exercise:** Consider a medium characterized entirely by a constant conductivity, σ. Demonstrate that the associated dielectric-permittivity function is

$$\varepsilon(\omega) = 1 - i\frac{\sigma}{\varepsilon_0\omega}. \tag{4.40}$$

4.4.3.1 Electrostatic fluctuations

For electrostatic or longitudinal oscillations, the current exactly cancels out Maxwell's displacement current, the wave-vector is parallel to the electric field, $k \parallel E$, and the fluctuations have no magnetic component. The electric field can be derived from an electric potential just like in electric circuits. In that case

$$\tilde{J}(k, \omega) = -i\omega\varepsilon_0\varepsilon(k, \omega)E(k, \omega), \tag{4.41}$$

and by the use of (4.24) the expression (4.25) results. It accounts for the thermal fluctuations of, for instance, longitudinal electrostatic waves such as Langmuir oscillations in plasmas (electron-plasma waves) and optical phonons in solids.

4.4.3.2 Electromagnetic fluctuations

For electromagnetic waves, assuming that the relative magnetic permeability $\mu_r = 1$, and that we have a scalar dielectric permittivity, elimination of H gives

$$\tilde{J}(k, \omega) = \frac{i}{\omega\mu_0}\left[k^2 - \left(\frac{\omega}{c}\right)^2\varepsilon(k, \omega)\right]E(k, \omega). \tag{4.42}$$

In this case the waves are *transverse*, with $k \perp E$. By using again the fluctuation–dissipation theorem (4.24) for the fluctuations in current, it is found from (4.42) that the transverse electromagnetic fluctuations in thermal equilibrium (Bekefi, 1966) have a power spectrum given by

$$\frac{1}{2}\varepsilon_0 E_{\text{eff}}^2(k; k + dk, \omega; \omega + d\omega) =$$
$$-\frac{\kappa T}{\pi}\frac{1}{\omega}\frac{\varepsilon_2(k, \omega)}{[\varepsilon_1(k, \omega) - k^2c^2/\omega^2]^2 + \varepsilon_2^2(k, \omega)}\frac{1}{(2\pi)^3}dk\,d\omega. \tag{4.43}$$

The corresponding results for the magnetic-field fluctuations are obtained by using Maxwell's equation and (4.43). For a vacuum, for which $\varepsilon_2(k, \omega) \to 0$ and $\varepsilon_1(k, \omega) \to 1$, the right-hand side of relation (4.43) approximates $(\kappa T/\omega)\delta(1 - k^2c^2/\omega^2)$, indicating that the fluctuations follow the dispersion relation $\omega^2 = c^2k^2$ for electromagnetic waves, as expected. Integration over all wavenumbers, followed by simple calculations, gives (Bekefi, 1966)

$$\frac{1}{2}\varepsilon_0 E_{\text{eff}}^2(\omega; \omega + d\omega) = \frac{\kappa T\omega^2}{4\pi^2c^3}d\omega. \tag{4.44}$$

By adding the analogous expression for the magnetic-field fluctuations, $\frac{1}{2}B^2(\omega)/\mu_0$, the standard expression for the black-body radiation of the electromagnetic wave's energy density at the temperature T is obtained, in the classical limit (Bekefi, 1966).

On Fourier transforming a one-dimensional periodic signal that is defined over an interval of unit length, we find that the density of discrete points along the wavenumber axis is $1/(2\pi)$. Similarly, for three-dimensional space, the density of wave modes in k-space is $1/(2\pi)^3$ for a unit volume. The number of wave modes in a frequency interval $\{\omega, \omega + d\omega\}$ is determined by the number of modes in a spherical shell in k-space with radius ω/c and thickness $dk = d\omega/c$, giving a volume of $4\pi(\omega^2/c^3)\,d\omega$, which, multiplied by the density of points, gives the number of modes $[\omega^3/(2\pi^2 c^3)]\,d\omega$. Consequently, the average energy per wave mode is κT; again we have equipartition of energy.

The fluctuation–dissipation theorem and its generalizations are relevant for describing many different physical processes. It has been applied, for example, to sound fluctuations (Callen and Welton, 1951), to molecular attractions in solids (Landau and Lifshitz, 1960), and to magnetic-resonance absorption (Kubo and Tomita, 1954). Applications to plasmas have been worked out by Thompson and Hubbard (1960), Hubbard (1961), Taylor (1960), Rostoker (1961), and Bekefi (1966). Density fluctuations in the ionospheric plasma allow the basic plasma parameters to be determined by measuring radar scattering, i.e. by remote sensing rather than the more complicated *in situ* measurements obtained with instrumented rockets (Bekefi, 1966). Applications specifically oriented toward radiation problems will be found in the works of Leontovich and Rytov (1952), Bunkin (1957), Haus (1961), and Mercier (1964). Fluctuations in solids are discussed by Kogan (1996).

The fluctuation–dissipation theorem discussed in this and the previous sections was obtained by considering systems having characteristics independent of the amplitude of the fluctuations, i.e. *linear* systems, as was emphasized before. This is not a trivial restriction.

4.5   Summary

A generalization of the fluctuation–dissipation theorem discussed in this chapter for special cases is achieved by replacing the electric force eE by a general force $\mathbf{F}$, the response of the current to the electric field is generalized to the time derivative of a response or 'displacement' ξ, and the dielectric-permittivity function is replaced by a response function χ. In summary

$$eE(k, \omega) \rightarrow F(k, \omega),$$
$$I(k, \omega) \rightarrow -i\omega\xi(k, \omega),$$
$$\varepsilon_0\varepsilon(k, \omega) \rightarrow \chi(k, \omega).$$

The derivation of the fluctuation–dissipation theorem outlined before made it natural to use a formulation in terms of power spectra. By use of the Wiener–Khinchine theorem, the fluctuation–dissipation theorem can also be formulated in terms of correlation functions, a form that is often used (Martin, 1968). Because of its fundamental importance, the theorem is repeated here once more in this alternative (classical) form, expressed in terms of the correlation function for the 'displacement' ξ and the imaginary part of the Fourier transform of the response function χ. With subscripts eq and neq denoting *equilibrium* and *nonequilibrium*, respectively, we have the following.

The dissipation that results when an external field is applied to a system is simply related to the fluctuations in thermodynamic equilibrium by

$$\langle \xi(t)\xi(t') \rangle_{\mathrm{eq}} - \langle \xi(t) \rangle_{\mathrm{eq}} \langle \xi(t') \rangle_{\mathrm{eq}} =$$
$$-\frac{2}{\pi} \kappa T \int_{-\infty}^{\infty} \frac{\chi_2(\omega)}{\omega} e^{i\omega(t-t')} d\omega,$$

where the response function (to be more specific, the *retarded* response function) relates the displacement to an external force $F^{\mathrm{ext}}(t)$ by

$$\langle \xi(t) \rangle_{\mathrm{neq}} = \int_{-\infty}^{t} \chi(t-t')F^{\mathrm{ext}}(t') dt'.$$

5   The Kramers–Kronig relations

The fluctuation–dissipation theorem relates thermal fluctuations to the dielectric function of a medium. Apparently, the fluctuations vanish in the limit $\varepsilon_2 \to 0$ for all frequencies, provided that this transition can be performed without affecting the real part of the dielectric function ε_1. A specific example demonstrated that in reality things need not be as simple as that. It turns out that there are general relations, the *Kramers–Kronig relations*, between the real and the imaginary parts of dielectric functions, relations that are actually independent of the assumption of thermal equilibrium. The theorem is connected to the much weaker assumption of causality.

A causal function is one whereby the response to an excitation depends only on the past, not on the future. For definiteness we here consider a Volterra-type, nonlocal, relation between the electric displacement $D(r, t)$ and the electric field $E(r, t)$, for which also the possibility of a nonlocal spatial dependence is allowed:

$$D(r, t) = p_0 E(r, t) + \iiint_{-\infty}^{\infty} \int_{-\infty}^{t} p(r - \xi, t - \tau) E(\xi, \tau) \, d\tau \, d^3\xi. \tag{5.1}$$

The first term accounts for the instantaneous response, with $p_0 = \text{constant}$. More generally, p can be taken to be a tensor in the case of anisotropic media. The electric displacement at a position r at a time t depends generally not only on the electric field at that position at that time, but also on E at neighboring positions with a spatially varying weight given through p. In addition, the actual value of D depends in general also on the 'history' of the electric field in the medium. The special case of a medium with temporal, but no spatial, dispersion corresponds to $p(r - \xi, t - \tau) = \delta(r - \xi)p(t - \tau)$. In this simple case the response is local in space, reducing (5.1) to the simpler form

$$D(t) = p_0 E(t) + \int_0^{\infty} p(\tau) E(t - \tau) \, d\tau,$$

where now the spatial variable is redundant, and therefore has been omitted.

Fourier transformation with respect to the spatial variable of the general formulation (5.1) gives

$$D(k, t) = p_0 E(k, t) + \int_{-\infty}^{t} p(k, t - \tau) E(k, \tau) \, d\tau. \tag{5.2}$$

The causality of the response is expressed by the upper limit t of integration in (5.2). In other words $D(k, t)$ depends only on the past, i.e. on $E(k, t)$ for times prior to the observation time t. The *spatial* variable does not give rise to similar considerations, so k enters only as a label in the following. For simple media, without spatial dispersion, the wavenumber dependence is trivial and can be omitted altogether. Fourier transformation with respect also to the temporal variable gives

$$D(k, \omega) = E(k, \omega)\left(p_0 + \int_0^\infty p(k, \tau)e^{-i\omega\tau}d\tau\right)$$

$$\equiv E(k, \omega)\varepsilon_0[1 + \chi(k, \omega)], \tag{5.3}$$

where now the susceptibility $\chi(k, \omega) = \int_0^\infty p(k, \tau)\exp(-i\omega\tau)\,d\tau$ of the medium is introduced. The dielectric function is defined as $\varepsilon(k, \omega) = 1 + \chi(k, \omega)$. The constant ε_0 is determined by requiring that $\chi(k, \omega) \to 0$ for $\omega \to \infty$. Physically, we expect all media to be transparent, or vacuum-like, for large frequencies. In particular, the constant ε_0 will be the response function for a vacuum, where $D = \varepsilon_0 E$. As an alternative notation one might see ε_0 included in $\varepsilon(k, \omega)$. A relation of the form (5.1) can be written for any spatially and temporally varying 'cause' (here E) and the corresponding linear 'response' (here D), $\varepsilon(k, \omega)$ being the Fourier transform of the transfer function.

The function $\varepsilon(k, \omega)$ is in general complex, with real and imaginary parts $\varepsilon_1(k, \omega)$ and $\varepsilon_2(k, \omega)$, respectively, and similarly for $\chi(k, \omega)$. The real and imaginary parts are, however, interrelated; we can not just take any two functions, one real and one imaginary, add them together, and call it a dielectric function! In order to demonstrate the relation between ε_1 and ε_2, it is advantageous first to rewrite (5.3). We introduce a pair of functions p_e and p_u, which are even and odd functions of time, respectively, defined such that $p(k, t) = p_e(k, t) + p_u(k, t)$ and $p(k, t) = 2p_e(k, t) = 2p_u(k, t)$ for $t > 0$, while $p(k, t) = 0$ for $t < 0$. On introducing Heaviside's unit step function $H(t) = 1$ for $t > 0$ and $H = 0$ for $t < 0$, we can write $p(k, t) = 2p_e(k, t)H(t)$. This expression is inserted as the integrand in (5.3) and the limits of integration are subsequently extended to $\{-\infty, \infty\}$.

In order to evaluate the Fourier transform of $p(k, \tau)$, as expressed before, we first recall that the Fourier transform of a product of two functions is the convolution of their individual transforms divided by 2π. The transform of $p(k, \tau)$ is by assumption $\chi(k, \omega) \equiv \chi_1(k, \omega) + i\chi_2(k, \omega)$.

The transform of the step function is obtained by considering $H(t)$ as the limiting form, for $\alpha \to 0$, of a function that is 0 for $t < 0$ and $\exp(-\alpha t)$ for $t > 0$, with $\alpha > 0$. The Fourier transform (Champeney, 1973) of this function is $(\alpha - i\omega)/(\alpha^2 + \omega^2)$, which has the limiting form $\pi\delta(\omega) - iP/\omega$ as $\alpha \to 0$. The letter P serves as a reminder that the principal value of the integral should be chosen when integrating with respect to ω. This can be made evident by considering the expression for a small, but nonzero, value of α, for which the singularity in an integration with respect to ω vanishes.

It is easily verified that the transform of p_e is $\chi_1(k, \omega)$, whereas p_u transforms into $i\chi_2(k, \omega)$. On performing the convolution of the two Fourier transforms, $\chi_1(k, \omega)$ and $\pi\delta(\omega) - iP/\omega$, the final result (Champeney, 1973) is then obtained as

$$\chi(k, \omega) = \chi_1(k, \omega) - \frac{i}{\pi}P\int_{-\infty}^\infty \frac{\chi_1(k, \omega_1)}{\omega - \omega_1}d\omega_1. \tag{5.4}$$

Alternatively, one might just as well use the relation $p(k, t) = 2p_u(k, t)H(t)$ and repeat the foregoing calculations to obtain

$$\chi(k, \omega) = i\chi_2(k, \omega) + \frac{1}{\pi}P\int_{-\infty}^\infty \frac{\chi_2(k, \omega_1)}{\omega - \omega_1}d\omega_1. \tag{5.5}$$

On comparing the two results, the following relations are readily confirmed.

$$\chi_2(\boldsymbol{k}, \omega) = -\frac{1}{\pi} P \int_{-\infty}^{\infty} \frac{\chi_1(\boldsymbol{k}, \omega_1)}{\omega - \omega_1} d\omega_1,$$

$$\chi_1(\boldsymbol{k}, \omega) = \frac{1}{\pi} P \int_{-\infty}^{\infty} \frac{\chi_2(\boldsymbol{k}, \omega_1)}{\omega - \omega_1} d\omega_1.$$

These are known as the Kramers–Kronig relations.

As indicated, these results are quite general, being valid for any stable, physically acceptable, causal transfer function. In the context of servo-mechanisms and electric circuits in general, they are often called the Bode relations (Bode, 1956). The integral transforms of the type used here are known as Hilbert transforms. We require χ_1 as well as χ_2 to be Hölder continuous for the principal values of the integrals to exist. This restriction is of little consequence for most practical problems.

The Kramers–Kronig relations are interesting by virtue of their introducing strict relationships concerning physical quantities solely on the basis of properties of complex functions; in this chapter, the electric susceptibility. The relations will evidently have implications for *any* causal response function. The real and imaginary parts of, say, a dielectric function are not independent, and, given one of them, the other can be determined by taking a Hilbert transform. Examples of functions forming a Hilbert-transform pair are shown in Figs. 5.2 and 5.4 later.

- **Exercise:** A function $f(x, y, z)$ is said to be Hölder continuous in the vicinity of a point ξ, η, ζ provided that a set of positive numbers h, α, and H can be found such that

$$R = \sqrt{(x - \xi)^2 + (y - \eta)^2 + (z - \zeta)^2} \le h$$

  implies $|f(x, y, z) - f(\xi, \eta, \zeta)| \le H R^\alpha$. A Hölder-continuous function is evidently continuous in the traditional sense. Demonstrate that a continuous function need not be Hölder continuous by considering $1/\ln(|x|)$ in the vicinity, but not *at* the singularlity in the origin, $x = 0$.

- **Exercise:** Prove the Kramers–Kronig relations by integrating the function $\chi(\boldsymbol{k}, \omega)/(\omega_0 - \omega)$ along the path shown in Fig. 5.1, letting $R \to \infty$ and $r \to 0$, and subsequently take the real and imaginary parts of the result. This proof is actually the standard one in the literature (Landau and Lifshitz, 1960). The one used in the text may make the relation to causality more apparent. Hint: demonstrate by use of (5.2) and (5.3) that $\chi(\boldsymbol{k}, \omega)$ is regular, i.e. without singularities, in the lower half of the complex ω-plane. (In the upper half of the plane the definition of $\chi(\boldsymbol{k}, \omega)$ has to be extended by analytic continuation, and in general it *does* have singularities there.)

- **Exercise:** Discuss the relation (4.40) in the light of the Kramers–Kronig relations. In particular, how should the integration contour in Fig. 5.1 be modified in order to accommodate the singularity at $\omega = 0$ in the imaginary part (4.40)? What is the *physical* origin of this singularity? Consider also (3.29) and (3.30).

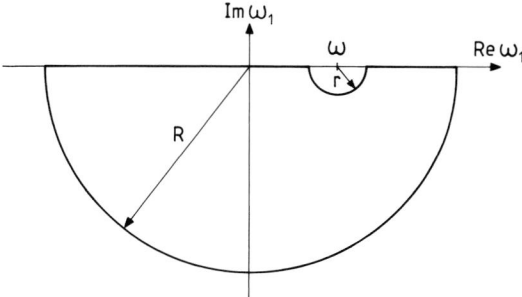

Figure 5.1 The integration path for the standard proof of the Kramers–Kronig relations, with $R \to \infty$ and $r \to 0$.

- **Exercise:** Demonstrate that the Kramers–Kronig relations can also (Champeney, 1973) be written as

$$\chi_2(\mathbf{k}, \omega) = -\frac{2}{\pi} \int_0^\infty \int_0^\infty \chi_1(\mathbf{k}, \omega_1) \sin(\omega t) \cos(\omega_1 t) \, d\omega_1 \, dt, \tag{5.6}$$

$$\chi_1(\mathbf{k}, \omega) = -\frac{2}{\pi} \int_0^\infty \int_0^\infty \chi_2(\mathbf{k}, \omega_1) \cos(\omega t) \sin(\omega_1 t) \, d\omega_1 \, dt. \tag{5.7}$$

- **Exercise:** By use of the Kramers–Kronig relations, prove the identity

$$P \int_{-\infty}^\infty \frac{d\omega_2}{(\omega - \omega_2)(\omega_2 - \omega_1)} = -\pi^2 \delta(\omega - \omega_1),$$

  known as the *Poincaré–Bertrand lemma* or one version of the *Plemelj formula*, see for instance van Kampen and Felderhof (1967).

- **Example:** Consider the case in which $\chi_1 = 0$. Then also $\chi_2 = 0$ for all ω, i.e. a dispersionless medium can not be dissipative and vice versa. The vacuum is such an example, but probably also the only one.

- **Example:** Consider the case in which $\chi_2 = -\delta(\omega - \omega_0)$, i.e. we have dissipation for one frequency only. Then $\chi_1 = (1/\pi)(\omega_0 - \omega)^{-1}$, implying that we have a resonance for ω_0. The response is *in* phase with the external forcing for frequencies below the resonance frequency, and in *counter phase* for frequencies $\omega > \omega_0$.

- **Example:** The basic equation describing the dynamics of collisionless, hot, and dilute plasmas is the Vlasov equation (Chen, 1983)

$$\frac{\partial f}{\partial t} + \mathbf{u} \cdot \nabla f + \frac{\mathbf{K}}{m} \cdot \nabla_u f = 0, \tag{5.8}$$

  where $f = f(\mathbf{r}, \mathbf{u}, t)$ is the velocity distribution of plasma particles with mass m, moving under the action of a force field $\mathbf{K} = \mathbf{K}(\mathbf{r}, \mathbf{u}, t)$. By ∇_u we understand the vector operator $\{\partial/\partial u_x, \partial/\partial u_y, \partial/\partial u_z\}$, in terms of derivatives with respect to velocity components.

For most relevant cases we will be dealing with electric and magnetic forces acting on individual particles with charge q and velocity $\boldsymbol{u}$, i.e. $\boldsymbol{K}(\boldsymbol{r}, \boldsymbol{u}, t) = q(\boldsymbol{E}(\boldsymbol{r}, t) + \boldsymbol{u} \times \boldsymbol{B}(\boldsymbol{r}, t))$ in terms of the Lorentz force. The electric and magnetic fields have to be determined by use of Maxwell's equations, whereby charge and current densities, $\rho(\boldsymbol{r}, t) \equiv q \int_{-\infty}^{\infty} f(\boldsymbol{r}, \boldsymbol{u}, t) \, d\boldsymbol{u}$ and $\boldsymbol{J}(\boldsymbol{r}, t) \equiv q \int_{-\infty}^{\infty} \boldsymbol{u} f(\boldsymbol{r}, \boldsymbol{u}, t) \, d\boldsymbol{u}$, respectively, are obtained self-consistently from the Vlasov equation; this is a highly complicated nonlinear problem.

To be explicit, we consider the case of high-frequency electrostatic electron-plasma waves in a collisionless plasma, where the ion component can be treated as an immobile background of uniformly distributed positive charge, and assume the force density $\boldsymbol{K}(\boldsymbol{r}, t) = -e\boldsymbol{E}(\boldsymbol{r}, t)$ in (5.8), for electrons with charge $-e$ and mass m.

It is readily seen that, with $\boldsymbol{K} = 0$, any function $f_0 = n_0 f_0(\boldsymbol{u})$, with $n_0 =$ constant, is a steady-state solution of (5.8). For convenience we use the normalization $\int_{-\infty}^{\infty} f_0(\boldsymbol{u}) \, d^3u = 1$. In thermal equilibrium, in particular, a spatially uniform plasma with a Maxwellian velocity distribution will be such a steady-state solution. Assume that we start out with the distribution f_0 and perturb it slightly at $t = t_0$ so that we have $f = f_0 + f_1$ and $n = n_0 + n_1$, with $n_1 \ll n_0$. For small-amplitude fluctuations, (5.8) can be linearized, i.e. second-order terms $\boldsymbol{E}_1 \cdot \nabla_u f_1$ can be ignored. By integrating the linearized version of (5.8) along characteristics, we obtain the solution

$$f_1(\boldsymbol{r}, \boldsymbol{u}, t) = f_1(\boldsymbol{r} - \boldsymbol{u}(t - t_0), \boldsymbol{u}, t_0) + \frac{en_0}{m} \int_{t_0}^{t} \boldsymbol{E}_1(\boldsymbol{r} - \boldsymbol{u}(t - t'), t') \cdot \nabla_u f_0(\boldsymbol{u}) \, dt'. \tag{5.9}$$

Note that (5.9) is not merely a rewriting of the linearized Vlasov equation; we have made a choice by writing a causal response to a given initial condition at $t = t_0$.

By integrating (5.9) with respect to the velocity $\boldsymbol{u}$ we obtain the electron density $n_1(\boldsymbol{r}, t)$, and find that

$$n_1(\boldsymbol{r}, t) = \int_{-\infty}^{\infty} f_1(\boldsymbol{r} - \boldsymbol{u}(t - t_0), \boldsymbol{u}, t_0) \, d^3u$$
$$+ \frac{en_0}{m} \int_{-\infty}^{\infty} \int_{t_0}^{t} \boldsymbol{E}_1(\boldsymbol{r} - \boldsymbol{u}(t - t'), t') \cdot \nabla_u f_0(\boldsymbol{u}) \, dt' \, d^3u. \tag{5.10}$$

For any nonsingular initial condition $f_1(\boldsymbol{r}, \boldsymbol{u}, t_0)$, the first term on the right-hand side of (5.10) vanishes by virtue of the 'phase mixing' at large times, and this contribution will be ignored in the following. In the same limit we let $t_0 \to -\infty$ in the integral, without loss of generality.

Since the functions $e^{-i\boldsymbol{k}\cdot\boldsymbol{r}}$ form a complete orthogonal set, we can expand the spatial variations of the potential and density by a Fourier transform, and consider only one arbitrary spatial Fourier component denoted by $\boldsymbol{k}$. For simplicity, we take $\boldsymbol{k}$ to be in the x-direction. With the assumption of electrostatic fluctuations, $\boldsymbol{E}_1 = -\nabla\phi$, we have $\boldsymbol{E}_1(\boldsymbol{r}, t) \to i\boldsymbol{k}\phi_1(\boldsymbol{k}, t)e^{-i\boldsymbol{k}\cdot\boldsymbol{r}}$. In this case the integrations of velocity components in the directions perpendicular to $\boldsymbol{k}$ are trivial, and we are left with a formally one-dimensional relation of the form

$$n_1(k, t)e^{-ikx} = \frac{en_0}{m} ik \int_{-\infty}^{\infty} \int_{-\infty}^{t} \phi_1(k, t')e^{-ik[x-u_x(t-t')]} \frac{d}{du_x} F(u_x) \, dt' \, du_x, \tag{5.11}$$

where the reduced distribution function F remains after integration with respect to u_y and u_z. For simplicity we write $F'(u_x) \equiv dF(u_x)/du_x$ in the following. The density is now inserted into Poisson's equation $k^2\phi_1(k, t) = -en_1(k, t)/\varepsilon_0$, as is appropriate for the particular Fourier component, giving the result

$$k^2\phi_1(k, t) = -\frac{e^2 n_0}{m\varepsilon_0} ik \int_{-\infty}^{\infty} \int_{-\infty}^{t} \phi_1(k, t')e^{iku_x(t-t')} F'(u_x) \, dt' \, du_x. \tag{5.12}$$

This result is now trivially rewritten in terms of Heaviside's step function, $H(\tau)$, and by introducing the new variable $\tau \equiv t - t'$, giving

$$k^2\phi_1(k, t) = -\frac{e^2 n_0}{m\varepsilon_0} ik \int_{-\infty}^{\infty} F'(v_x) \int_{-\infty}^{\infty} H(\tau)e^{iku_x\tau}\phi_1(k, t - \tau) \, d\tau \, du_x. \tag{5.13}$$

On Fourier transforming with respect to time, we note that the right-hand side of (5.13) contains a convolution of $H(\tau)e^{iku_x\tau}$ and $\phi_1(k, t - \tau)$. The Fourier transform of this convolution is expressed by the product of the Fourier transforms of $H(\tau)e^{iku_x\tau}$ and $\phi_1(k, \tau)$. We denote the transform of the potential by $\phi_1(k, \omega)$. The Fourier transform of the product $H(\tau)e^{iku_x\tau}$ is the transform of $H(\tau)$ shifted by $-ku_x$, giving $\pi\delta(\omega - ku_x) - iP/(\omega - ku_x)$, with P denoting that the principal value has to be taken upon integration, as before. Recall that $\delta(ax) = \delta(x)/|a|$, see Appendix D, so we will take $k > 0$ in the following. By insertion we find that now the transforms of the electrostatic potential cancel, and we are left with the dispersion relation in the form

$$1 - \frac{\omega_{pe}^2}{k^2} P \int_{-\infty}^{\infty} \frac{F'(u_x)}{u_x - \omega/k} \, du_x + i\pi\frac{\omega_{pe}^2}{k^2} F'(\omega/k) = 0, \tag{5.14}$$

where the electron-plasma frequency, ω_{pe}, has been introduced.

Note the relative ease with which this relation was obtained, compared with the complicated manipulations in complex planes by which the problem is usually approached by Laplace transforming an initial-value problem (Chen, 1983).

To obtain an analytical expression for the dielectric function associated with a plasma in its kinetic description, we first recall the general definition of the relative dielectric function of a dispersive medium

$$\varepsilon(\boldsymbol{k}, \omega) = 1 - \frac{en_1(\boldsymbol{k}, \omega)}{\varepsilon_0 k^2 \phi_1(\boldsymbol{k}, \omega)}, \tag{5.15}$$

where again only one spatial Fourier component was considered. With the previous arguments we find

$$n_1(k, t) = \frac{en_0}{m} ik \int_{-\infty}^{\infty} F'(u_x) \int_{-\infty}^{\infty} H(\tau)e^{iku_x\tau}\phi_1(k, t - \tau) \, d\tau \, du_x, \tag{5.16}$$

to be used in (5.15). By Fourier transforming (5.16) with respect to time as before, we find the relative dielectric function in the form

$$\varepsilon(k, \omega) = 1 - \frac{\omega_{pe}^2}{k^2} P \int_{-\infty}^{\infty} \frac{F'(u_x)}{u_x - \omega/k} \, du_x + i\pi\frac{\omega_{pe}^2}{k^2} F'(\omega/k), \tag{5.17}$$

for real frequencies ω. The full dielectric function is $\varepsilon_0\varepsilon(k, \omega)$. Recall that $F'(\omega/k) < 0$ for a Maxwellian distribution with $\omega/k > 0$ as has implicitly been assumed, implying that $\varepsilon_2(k, \omega) < 0$, as is expected in thermal equilibrium. (With other sign conventions than the one used here, we have a change in sign of the imaginary part.) As expected, we find the dispersion relation for electrostatic waves by requiring that $\varepsilon(k, \omega) = 0$.

Note that the Vlasov equation (5.8) is *time reversible* (verify this for K given as the full Lorentz force!). The imaginary part of the dielectric function (5.17) is due to Landau damping (Chen, 1983). In thermal equilibrium, it is found that, for a plane wave $\exp[i(\omega t - kx)]$, the scalar dielectric function of plasmas can be expressed as

$$\varepsilon(k, \omega) = 1 - \frac{1}{2(k\lambda_D)^2} Z'\left(\frac{\omega}{k}\sqrt{\frac{m}{2\kappa T}}\right) \tag{5.18}$$

where m is the electron mass, T is the electron temperature, λ_0 is the Debye length, and Z' is the plasma's dispersion function given as the derivative of the Z-function (Fried and Conte, 1961)

$$Z(u) = -2ie^{-u^2}\int_0^{iu} e^{-\xi^2}\,d\xi - i\sqrt{\pi}e^{-u^2}$$

$$= -2e^{-u^2}\int_0^u e^{\gamma^2}\,d\gamma - i\sqrt{\pi}e^{-u^2}, \tag{5.19}$$

with u real. Evidently, Z is closely related to the error function of the complex argument. We have $Z'(u) = -2[1 + uZ(u)]$. For plane waves $\exp[-i(\omega t - kx)]$, we use the complex conjugate of (5.19). The susceptibility of the plasma obtained from (5.18) is evidently $\chi(k, \omega) = -\frac{1}{2}Z'\left[(\omega/k)\sqrt{m/(2\kappa T)}\right]/(k\lambda_D)^2$, which is shown in Fig. 5.2, apart from a coefficient $1/(k\lambda_D)^2$. For comparison see (3.29) and (3.30) describing collisional damping of Langmuir oscillations.

In deriving (5.18), it was assumed that the ions are forming a neutralizing background of positive charge, arguing that the frequency of the electron plasma waves is so high that inertia renders the ions effectively immobile.

- **Exercise:** Demonstrate that the real and imaginary parts of the derivative Z' of the plasma's dispersion function (5.19) indeed form a Hilbert-transform pair.

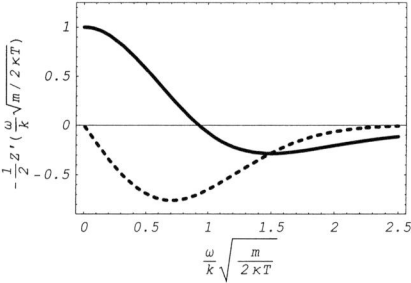

Figure 5.2 Real (full line) and imaginary (dashed line) parts of the plasma's susceptibility function $-\frac{1}{2}Z'$, for real arguments.

The condition that D should be real for real E imposes the 'reality condition' $\varepsilon(-\boldsymbol{k}, -\omega) = \varepsilon^*(\boldsymbol{k}, \omega)$, with real ω and $\boldsymbol{k}$, as is obtained for instance from the relations (5.1) and (5.3). The reality condition implies that $\varepsilon_1(-\boldsymbol{k}, -\omega) = \varepsilon_1(\boldsymbol{k}, \omega)$ and $\varepsilon_2(-\boldsymbol{k}, -\omega) = -\varepsilon_2(\boldsymbol{k}, \omega)$. Taking these properties into account, the Kramers–Kronig relations can be written as

$$\chi_2(\boldsymbol{k}, \omega) + \chi_2(-\boldsymbol{k}, \omega) = -\frac{2\omega}{\pi} P \int_0^\infty \frac{\chi_1(\boldsymbol{k}, \omega_1) + \chi_1(-\boldsymbol{k}, \omega_1)}{\omega^2 - \omega_1^2} \, d\omega_1, \tag{5.20}$$

$$\chi_1(\boldsymbol{k}, \omega) + \chi_1(-\boldsymbol{k}, \omega) = \frac{2}{\pi} P \int_0^\infty \frac{\omega_1 [\chi_2(\boldsymbol{k}, \omega_1) + \chi_2(-\boldsymbol{k}, \omega_1)]}{\omega^2 - \omega_1^2} \, d\omega_1. \tag{5.21}$$

These results become particularly simple for $\boldsymbol{k} = 0$. To simplify the notation, only $\chi(\boldsymbol{k} = 0, \omega)$ will be considered in the next section. For simplicity we write $\chi(\omega)$ for $\chi(0, \omega)$.

- **Example:** Dielectric media can support so-called electrostatic waves, i.e. waves with fluctuating electric fields, but no associated magnetic-field variations, optical phonons and Langmuir oscillations in plasmas being two examples. For the magnetic field to vanish, it is required that the current fluctuations, $\tilde{\boldsymbol{J}}$, exactly cancel out Maxwell's displacement current, $\partial \boldsymbol{D}/\partial t$, in (4.26). In order to have freely propagating waves, i.e. fluctuating electric fields that are not maintained by external charges, we require $\varepsilon(\boldsymbol{k}, \omega) = 0$ in (4.33) or (4.38). (The term 'electrostatic' is a misnomer, but well established in the literature. There is nothing 'static' about these waves; it is just that their properties are derived from the same equations as those used for electrostatics.)

  The dielectric losses will give rise to damping of waves. The damping rate is determined by the real as well as the imaginary part of $\varepsilon(\boldsymbol{k}, \omega)$. Assume that the medium supports weakly damped waves with the dispersion relation $\omega(\boldsymbol{k}) = \omega_1(\boldsymbol{k}) + i\omega_2(\boldsymbol{k})$ with $\omega_1(\boldsymbol{k}) \gg \omega_2(\boldsymbol{k})$. These waves are associated with a zero of the dielectric function $\varepsilon(\boldsymbol{k}, \omega)$ as discussed before. For weakly damped waves, we look for zeros close to the real ω-axis, and approximate

  $$\varepsilon(\boldsymbol{k}, \omega) \equiv \varepsilon_1(\boldsymbol{k}, \omega_1 + i\omega_2) + i\varepsilon_2(\boldsymbol{k}, \omega_1 + i\omega_2)$$
  $$\approx \varepsilon_1(\boldsymbol{k}, \omega_1) + i\varepsilon_2(\boldsymbol{k}, \omega_1) - \omega_2 \frac{\partial \varepsilon_2}{\partial \omega_1} + i\omega_2 \frac{\partial \varepsilon_1}{\partial \omega_1}. \tag{5.22}$$

  We assume that the third term is small, because both ω_2 and ε_2 were assumed small. The condition $\varepsilon(\boldsymbol{k}, \omega) = 0$ then implies that $\varepsilon_1(\boldsymbol{k}, \omega_1) = 0$, which determines the real part of the dispersion relation for the waves, $\omega_1 = \omega_1(\boldsymbol{k})$. The imaginary part of (5.22) gives

  $$\omega_2(\boldsymbol{k}) = -\varepsilon_2(\boldsymbol{k}, \omega_1(\boldsymbol{k})) \left/ \frac{\partial \varepsilon_1(\boldsymbol{k}, \omega)}{\partial \omega} \right|_{\omega = \omega_1(\boldsymbol{k})}. \tag{5.23}$$

  The numerator is given by the dielectric losses, while the denominator is determined by the real part of $\varepsilon(\boldsymbol{k}, \omega)$. For damping of a wave, we require $\omega_2(\boldsymbol{k}) > 0$ with the assumed time variation $\exp(i\omega t)$. Since $\varepsilon_2(\boldsymbol{k}, \omega_1(\boldsymbol{k})) < 0$ in thermal equilibrium, this implies that $\partial \varepsilon_1(\boldsymbol{k}, \omega)/\partial \omega > 0$ at a frequency $\omega_1(\boldsymbol{k})$ at which $\varepsilon_1(\boldsymbol{k}, \omega_1) = 0$.

  The foregoing discussion implicitly concerned *temporal* wave damping; we imagined an initial condition such that a plane wave was released and we subsequently followed its damping in time. Physically, the alternative situation is more interesting; a harmonic wave is excited at a certain spatial position by an antenna, and we follow the damping of the wave as it propagates away from the exciter. Assume that ω is real, while $k = k_1 + ik_2$, and use the expansion

$$\varepsilon(k, \omega) \approx \varepsilon_1(k_1, \omega) + i\varepsilon_2(k_1, \omega) - k_2 \cdot \nabla_k\varepsilon_2(k, \omega) + ik_2 \cdot \nabla_k\varepsilon_1(k, \omega).$$
$$(5.24)$$

As before we obtain $\varepsilon_1(k_1, \omega) = 0$ determining $\omega = \omega(k_1)$, which is essentially the same result as before, but now the damping is determined by

$$k_2 \cdot \nabla_k\varepsilon_1(k_1, \omega) = -\varepsilon_2(k_1, \omega(k_1)),$$
$$(5.25)$$

where in general $\nabla_k\varepsilon_1$ need not be parallel with k_2. In particular, we find

$$\omega_2 = -k_2 \cdot \nabla_k\omega_1$$
$$(5.26)$$

using

$$\nabla_k\omega_1 = -\nabla_k\varepsilon_1(k, \omega_1) \bigg/ \frac{\partial\varepsilon_1(k, \omega_1)}{\partial\omega_1}$$

for the group velocity.

Since the Kramers–Kronig relations determine $\varepsilon_1(k, \omega)$ for given $\varepsilon_2(k, \omega)$, and vice versa, we find that a change in the damping of an electrostatic wave will necessarily be accompanied by a change in its dispersion relation, i.e. its propagation velocity. Similar conclusions will of course hold for any other type of wave.

The thermal fluctuations described by (4.25) can be qualitatively understood as 'radiation' of electrostatic waves by the electrons in thermal motion; a selected electron (or ion, when low-frequency waves are considered also) gives rise to a radiation of electrostatic waves (like Čerenkov radiation of waves) in a dielectric medium composed by all the other electrons. This description is called the 'superposition of dressed particles' (Bekefi, 1966), the 'dress' being the polarization of the surrounding medium induced by the selected particle.

- Exercise: Try out the results (5.23) and (5.25) on the dielectric function (5.17) to obtain the expression for collisionless Landau damping of electron plasma waves.

5.1   Anisotropic media

The foregoing discussion implicitly assumed that the media were isotropic; the directions of, for instance, $E(k, \omega)$ and $D(k, \omega)$ were the same, $\varepsilon_0 \varepsilon(k, \omega)$ being a scalar proportionality factor. In a more general case, for anisotropic media, we have $D(k, \omega) = \epsilon(k, \omega) \cdot E(k, \omega)$, where $\epsilon(k, \omega)$ is a tensor with components $\epsilon_{j\ell}(k, \omega)$. The Kramers–Kronig relations will hold for the real and imaginary parts of each component.

- **Exercise:** Write the dielectric tensor in terms of its Hermitian and anti-Hermitian parts as $\epsilon_{j\ell}(k, \omega) = \epsilon_{j\ell}^{(')}(k, \omega) + i\epsilon_{j\ell}^{('')}(k, \omega)$, where $\epsilon_{j\ell}^{(')}(k, \omega) = \frac{1}{2}[\epsilon_{j\ell}(k, \omega) + \epsilon_{\ell j}^*(k, \omega)]$ and $\epsilon_{j\ell}^{('')}(k, \omega) = -(i/2)[\epsilon_{j\ell}(k, \omega) - \epsilon_{\ell j}^*(k, \omega)]$. Both $\epsilon_{j\ell}^{(')} = \epsilon_{\ell j}^{(')*}$ and $\epsilon_{j\ell}^{('')} = \epsilon_{\ell j}^{('')*}$ are Hermitian tensors, whereas $i\epsilon_{j\ell}^{('')}$ is anti-Hermitian. In particular, the diagonal elements $\epsilon_{jj}^{(')}$ and $\epsilon_{jj}^{('')}$ are real.

Write out the Kramers–Kronig relations for the Hermitian and anti-Hermitian parts of $\epsilon_{j\ell}(\mathbf{k}, \omega)$.

The 3×3 tensor $\epsilon_{j\ell}(\mathbf{k}, \omega)$ will have nine components. These are, however, not independent; it can be shown from some very general irreversible thermodynamic principles that, for any nonactive medium in an externally imposed static magnetic field $\mathbf{B}_0$, we have the Onsager relations (Yeh and Liu, 1972, Pathria, 1996)

$$\epsilon_{j\ell}(\mathbf{k}, \omega, \mathbf{B}_0) = \epsilon_{\ell j}(-\mathbf{k}, \omega, -\mathbf{B}_0).$$

If, in addition, the medium is nongyrotropic, we have $\epsilon_{j\ell}(\mathbf{k}, \omega, \mathbf{B}_0) = \epsilon_{\ell j}(-\mathbf{k}, \omega, \mathbf{B}_0)$. Then $\epsilon_{j\ell}(\mathbf{k}, \omega, \mathbf{B}_0) = \epsilon_{j\ell}(\mathbf{k}, \omega, -\mathbf{B}_0)$. The Onsager relation ultimately implies that $\epsilon_{j\ell}(\mathbf{k}, \omega)$ reduces to six independent elements. For a detailed discussion, we refer the reader to Yeh and Liu (1972) for instance.

If, in addition to the principle of causality, we also take into account that all information is propagated with a finite velocity in spatially dispersive media, then additional relations between the Hermitian and anti-Hermitian parts of the dielectric tensor can be derived.

5.2   Sum rules

For a passive system in thermal equilibrium we have dissipation at all frequencies, meaning that $\omega\chi_2(\omega) \leq 0$. In particular, we expect that $\chi_2 \to 0$ for $\omega \to \infty$, since no physical system can respond or dissipate at infinite frequencies, due to there being finite inertia. If, for instance, the forcing frequency of a pendulum increases to well above the resonance frequency, its displacement becomes vanishingly small. Provided that $\omega\chi_2(\omega) \to 0$ for $\omega \to \infty$, we can ignore ω_1^2 relative to ω^2 in this limit and obtain from (5.21)

$$\chi_1(\omega) \approx \frac{2}{\pi\omega^2} \int_0^\infty \omega_1 \chi_2(\omega_1) \, d\omega_1. \tag{5.27}$$

For large ω we thus have $\chi_1(\omega) < 0$ and varying approximately as ω^{-2}. For $\omega = 0$, on the other hand,

$$\chi_1(\omega = 0) = -\frac{2}{\pi} \int_0^\infty \frac{1}{\omega_1} \chi_2(\omega_1) \, d\omega_1. \tag{5.28}$$

Since $\chi_2(\omega) = -\chi_2(-\omega)$, we expect $\chi_2(\omega = 0) = 0$ in general, so there is no singularity in this integral. (Some nontrivial exceptions can be found, though, for which $\chi_2(\omega = 0)$ is singular, see Section 3.1. Such cases need a separate analysis.) Since $\chi_2(\omega) \leq 0$, the relation (5.28) implies that $\chi_1(\omega = 0) > 0$ in general. An asymptotic expansion of χ_1 can be obtained by use of $(\omega - \omega_1)^{-1} = \omega^{-1} \sum_n^\infty (\omega_1/\omega)^n$, giving

$$\chi_1(\omega) \approx \frac{2}{\pi\omega} \sum_{n=0}^\infty \int_0^\infty \chi_2(\omega_1) \left(\frac{\omega_1}{\omega}\right)^{2n+1} d\omega_1, \tag{5.29}$$

which reproduces the previous result (5.27) when we retain the term with $n = 0$ only. It is evidently assumed that the integral converges for all n, requiring that χ_2 decreases faster than any power of ω, e.g. exponentially.

A relation for the integral of χ_1 can also be obtained. Thus, for large ω we find from (5.20)

$$\frac{2}{\pi} \int_0^\infty \chi_1(\omega_1)\, d\omega_1 = -\lim_{\omega\to\infty} [\omega\chi_2(\omega)], \tag{5.30}$$

demonstrating that the integral of χ_1 depends solely on the properties of χ_2 at $\omega \to \infty$. The limit on the right-hand side is zero, for most relevant cases.

- **Exercise:** Write out the consequences of the reality conditions for the Kramers–Kronig relations for $k \neq 0$.

- **Exercise:** Prove that

$$\int_0^\infty \chi_1(\omega)\chi_2(\omega)\frac{1}{\omega}\, d\omega = \frac{\pi}{4}\chi_1^2(\omega = 0).$$

Hint: integrate the function $\chi_2(\omega)/\omega$ along a path similar to the one shown in Fig. 5.1 with a small half-circle around $\omega = 0$, and take the real and imaginary parts of the result. Use the result to prove that

$$\int_0^\infty \left(\chi_1(\omega) - \frac{1}{2}\chi_1(\omega = 0)\right)\chi_2(\omega)\, d\ln\omega = 0.$$

This is actually interesting: $\chi_1(\omega) - \frac{1}{2}\chi_1(\omega = 0)$ and $\chi_2(\omega)$ are orthogonal when measured on a logarithmic frequency scale, $\ln\omega$.

The sum rules given hitherto basically reflect the properties of the Hilbert transforms which are the essential constituents of the Kramers–Kronig relations. There are, however, also constraints that are imposed by basic physical principles (Martin, 1968). We consider here the implications of the fluctuation–dissipation theorem. By Parseval's theorem we have $(1/T)\int_0^T I^2(t)\, dt = \sum_l |I_l|^2$, and use of (4.17) gives

$$\frac{1}{T}\int_0^T \langle I^2(t)\rangle\, dt = \langle I^2(t)\rangle = -\frac{2}{\pi}\kappa T\frac{\varepsilon_0\mathcal{A}}{\mathcal{L}}\int_0^\infty \omega\varepsilon_2(\omega)\, d\omega, \tag{5.31}$$

since $\langle I^2(t)\rangle$ is independent of time for statistically time-stationary processes. Similarly, from (4.18) we have

$$\frac{1}{T}\int_0^T \langle V^2(t)\rangle\, dt = \langle V^2(t)\rangle$$

$$= -\frac{2}{\pi}\kappa T\frac{\mathcal{L}}{\varepsilon_0\mathcal{A}}\int_0^\infty \frac{1}{\omega}\frac{\varepsilon_2(\omega)}{\omega\varepsilon_1^2(\omega) + \varepsilon_2^2(\omega)}\, d\omega \tag{5.32}$$

for the fluctuating voltage associated with a condenser filled with a dielectric. The two integrals in (5.31) and (5.32) must be finite for physically acceptable media to keep the mean-square current and voltage fluctuations finite.

Similarly we expect that

$$\left\langle \left(\frac{d^n I}{dt^n}\right)^2\right\rangle = -\frac{2}{\pi}\kappa T\frac{\varepsilon_0\mathcal{A}}{\mathcal{L}}\int_0^\infty \omega^{2n+1}\varepsilon_2(\omega)\, d\omega \tag{5.33}$$

and

$$\left\langle \left(\frac{d^n V}{dt^n}\right)^2\right\rangle = -\frac{2}{\pi}\kappa T\frac{\mathcal{L}}{\varepsilon_0\mathcal{A}}\int_0^\infty \omega^{2n-1}\frac{\varepsilon_2(\omega)}{\varepsilon_1^2(\omega) + \varepsilon_2^2(\omega)}\, d\omega \tag{5.34}$$

must both be finite for all n for physically realizable fluctuating signals, assuming that such signals are differentiable infinitely many times. The consequence is that certain conditions are imposed on $\varepsilon_1(\omega)$ and $\varepsilon_2(\omega)$ as $\omega \to \infty$, see also (5.29). Many models will fail to satisfy these conditions, and it is instructive to locate the reason for their shortcoming (Martin, 1968).

- **Exercise:** Derive the electric polarization for a medium with immobile ions and harmonically bound electrons with a simple damping term, such that the average electron position in an externally applied electric field, E, follows

$$m \frac{d^2}{dt^2} \langle x \rangle + m\omega_0^2 \langle x \rangle + m\gamma \frac{d}{dt} \langle x \rangle = -eE, \tag{5.35}$$

  ω_0 being a natural oscillation of the system in the limit of vanishing damping coefficient, $\gamma \to 0$. Obtain the dielectric function for this simple-oscillator model (5.35), see also Section 3.4.1. Prove explicitly that the result satisfies the Kramers–Kronig relations. (For this simple model, you can solve the integrals analytically.) Will the result satisfy (5.33) and (5.34) for all n? Explain!

- **Exercise:** Try to improve on the simple-oscillator model by introducing a phenomenologic damping term $m[e^2/(6\pi\varepsilon_0 mc^3)]\, d^3 \langle x \rangle/dt^3$, which accounts for the radiation losses from accelerated electrons (Martin, 1968). How much of an improvement does such a term represent, seen in the light of (5.33) and (5.34)? The implications and limitations of the radiation term mentioned before are discussed for instance by Clemmow and Dougherty (1990).

- **Exercise:** Determine the asymptotic expansion of Z' in (5.19) and use it in (5.18) to recover the standard cold-plasma dielectric function $\varepsilon(\omega) = 1 - \omega_{pe}^2/\omega^2$.

5.2.1   More about the Kramers–Kronig relations

The foregoing discussion summarizes the basic properties of the Kramers–Kronig relations. There are, however, several additional expressions that can be useful sometimes. The basic idea is that a rather good freehand drawing of the real part of a response function can be produced when its imaginary part is known, and vice versa. The relevant relations are not proven here in detail; only an outline of the procedure is given.

First, using $P \int_{-\infty}^{\infty} (\omega^2 - \omega_1^2)^{-1} \, d\omega_1 = 0$, the relation (5.21), is rewritten in the form

$$\chi_1(\omega) = \frac{2}{\pi} \int_0^\infty \frac{\omega_1 \chi_1(\omega_1) - \omega\chi_1(\omega)}{\omega^2 - \omega_1^2} \, d\omega_1, \tag{5.36}$$

and the principal-value sign can now be omitted. There is no singularity in the integral at $\omega = \omega_1$. On introducing the substitutions $u = \ln(\omega_1/\omega)$, $\omega_1/\omega = \exp u$, and $d\omega_1/\omega_1 = du$, we obtain

$$\chi_1(\omega) = -\frac{1}{\pi} \int_{-\infty}^\infty \ln\left[\coth\left(\left|\frac{u}{2}\right|\right)\right] \frac{d[e^u \chi_2(u)]}{du} \, du. \tag{5.37}$$

It is readily seen that $\ln[\coth(|u/2|)]$ is strongly peaked at $u = 0$, and the integral obtains its main contribution around this value, provided that $e^u \chi_2(u)$ is gently varying. Using that $d[e^u \chi_2(u)]/du|_{u=0} = d[\omega \chi_2(\omega)]/d\omega$ and also $\int_{-\infty}^{\infty} \ln[\coth(|u/2|)]\, du = \pi^2/2$, we obtain

$$\chi_1(\omega) = -\frac{\pi}{2} \frac{d[\omega \chi_2(\omega)]}{d\omega} - \frac{1}{\pi} \int_{-\infty}^{\infty} \left(\frac{d[e^u \chi_2(u)]}{du} - \frac{d[\omega \chi_2(\omega)]}{d\omega} \right) \ln\left[\coth\left(\left|\frac{u}{2}\right| \right) \right] du. \quad (5.38)$$

Since, by construction, the term in the brackets in the integrand of (5.38) vanishes at the value of u at which $\ln[\coth(|u/2|)]$ is peaked, see Fig. 5.3, it can be argued that the integral is small, generally. One might consequently expect that $\chi_1(\omega) = -(\pi/2)\,d[\omega\chi_2(\omega)]/d\omega$ is a good approximation. Do not expect too good an agreement, though; after all, it is only an approximation!

With some manipulations, involving a contour integration of the function $\chi(\omega)/[\omega(\Omega - \omega_0)]$, an alternative expression for $\chi_1(\omega)$ can be obtained:

$$\chi_1(\omega) - \chi_1(\omega = 0) = -\frac{\pi}{2} \omega^2 \frac{d[\chi_2(\omega)/\omega]}{d\omega}$$
$$- \frac{1}{\pi} \int_{-\infty}^{\infty} \left(\frac{d[e^{-u}\chi_2(u)]}{du} - \omega^2 \frac{d[\chi_2(\omega)/\omega]}{d\omega} \right) \ln\left[\coth\left(\left|\frac{u}{2}\right| \right) \right] du, \quad (5.39)$$

giving an approximation for χ_1 that differs from (5.38). This could be expected, since, after all, the integral contributions in (5.38) and (5.39) are not exactly vanishing. The first approximation is best for large ω, the latter one for small ω.

The foregoing expressions assume that the imaginary part $\chi_2(\omega)$ is known. This is probably most often the case, for this function is associated with losses, which can be measured relatively easily. Similar relations can be obtained for use when $\chi_1(\omega)$ is given *a priori*. From (5.20) we find

$$\chi_2(\omega) = -\frac{2\omega}{\pi} \int_0^{\infty} \frac{\chi_1(\omega_1) - \chi_1(\omega)}{\omega^2 - \omega_1^2}\, d\omega_1. \quad (5.40)$$

With the substitutions used in (5.37), this relation is rewritten in terms of the variable u introduced before,

$$\chi_2(\omega) = \frac{\pi}{2} \omega \frac{d\chi_1(\omega)}{d\omega} + \frac{1}{\pi} \int_{-\infty}^{\infty} \left(\frac{d\chi_2(u)}{du} - \omega \frac{d\chi_2(\omega)}{d\omega} \right) \ln\left[\coth\left(\left|\frac{u}{2}\right| \right) \right] du. \quad (5.41)$$

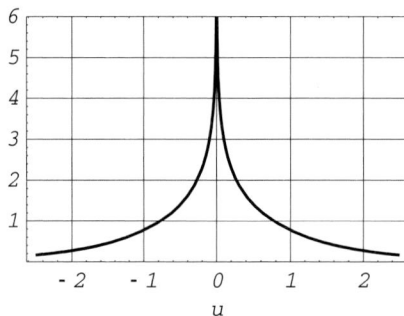

Figure 5.3 The function $\ln[\coth(|u/2|)]$.

With reference to the arguments used before, it might again be concluded that one can assume that $\chi_2(\omega)$ is well approximated by $(\pi/2)\omega\, d\chi_1(\omega)/d\omega$.

- **Exercise:** Try out the foregoing approximations on the simple-oscillator model, for which the exact relations are

$$\chi_1(\omega) = \frac{\omega_0^2 - \omega^2}{(\omega_0^2 - \omega^2)^2 + v^2\omega^2}, \tag{5.42}$$

$$\chi_2(\omega) = -\frac{v\omega}{(\omega_0^2 - \omega^2)^2 + v^2\omega^2}, \tag{5.43}$$

where ω_0 is a resonance frequency and v is a friction coefficient, see Fig. 5.4. Try to plot the response function for various ratios of v/ω_0. Derive the full expression for the time-varying response function $p(t)$ from $\chi(\omega)$. The oscillator model is the basic element in the Drude–Lorentz dielectric model.

- **Exercise:** Prove the relations

$$\chi_1(\omega) = -\frac{1}{\omega\pi} \int_0^\infty \frac{d[\omega_1\chi_2(\omega_1)]}{d\omega_1} \ln\left(\left|\frac{\omega_1 + \omega}{\omega_1 - \omega}\right|\right) d\omega_1, \tag{5.44}$$

$$\chi_2(\omega) = \frac{1}{\pi} \int_0^\infty \frac{d\chi_1(\omega_1)}{d\omega_1} \ln\left(\left|\frac{\omega_1 + \omega}{\omega_1 - \omega}\right|\right) d\omega_1. \tag{5.45}$$

Because we have $\omega\chi_2 < 0$, with the present convention for ω, the integral in (5.45) must be negative. Since the function $\ln[|(\omega_1 + \omega)/(\omega_1 - \omega)|]$ is sharply peaked at $\omega_1 = \omega$, see Fig. 5.5, this implies that χ_2 is likely to have its most negative value at a frequency at which $d\chi_1(\omega)/d\omega < 0$ and its minimum value (i.e. $\chi_2 \sim 0$) where $d\chi_1(\omega)/d\omega \sim 0$, in agreement also with (5.41).

Extensions of these approximations are outside the scope of the present analysis and the interested reader is referred to the work of Bode (1956).

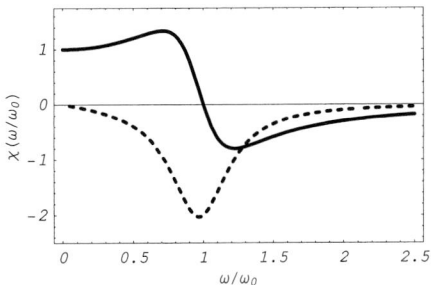

Figure 5.4 Real (full line) and imaginary (dashed line) parts of the response function $\chi(\omega)$ derived from the simple-oscillator model, e.g. (5.42) and (5.43), with $v/\omega_0 = 0.5$. We normalized the real part to unity at $\omega = 0$.

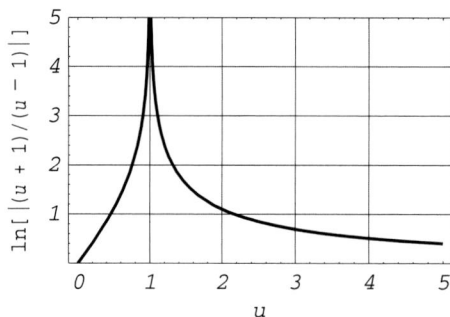

Figure 5.5 The function $\ln[|(u+1)/(u-1)|]$, where $u \equiv \omega_1/\omega$.

5.2.2   Comments on sign conventions

There seems to be a certain confusion concerning the convention for Fourier transforms. Sometimes $e^{-i\omega t}$ is used, sometimes $e^{i\omega t}$, corresponding to different signs of ω. In electric engineering there seems to be a preference for the $+$ sign and the letter i is replaced by j. Physicists, on the other hand, tend to use the $-$ sign. There is no logical resolution of this ambiguity since physically we can not discriminate between counter-clockwise and clockwise rotation in the complex plane, as mentioned previously. The consequence is an ambiguity in sign of $\chi_2(k, \omega)$, i.e. in the latter case it must be positive for dissipation. Fortunately there seems to be general agreement that frequencies read on a spectrum analyzer are positive!

Finally, sometimes the complex dielectric function is defined with a minus sign as $\varepsilon(k, \omega) = \varepsilon_1(k, \omega) - i\varepsilon_2(k, \omega)$. This again corresponds to a change in sign of the imaginary part ε_2.

6   **Brownian motion**

The origin of the thermal fluctuations in the simple model in Chapter 3 was identified as the thermal motion of the free electron in the circuit. A consequence of this motion was the current induced in the short-circuiting wire, and the statistical characteristics of this current were analyzed in detail. The displacement of the electron itself is also a random process that can be analyzed similarly. The random displacement of the electron and the fluctuations in current are two different manifestations of the same phenomenon. An essential difference between the two problems is, however, that, while the induced current represented a time-stationary process, this is not the case for the particle displacement, as will become evident from the following analysis.

6.1   Diffusion of microscopic particles

When analyzing the random displacement of small particles it is convenient to distinguish between two extreme limits. One is the case of microscopic particles, electrons or ions for instance, whereas the other limit is that of macroscopic particles such as pollen and dust particles. These are of course also small, but can be inspected directly, e.g. by using microscopes.

6.1.1   Diffusion of a light particle

The displacement of the point charge considered in Section 3.1 can be analyzed quite simply. Since the velocity is constant between collisions, its displacement in the z-direction is $z = z_0 + wt$ until the next collision. The displacement with respect to z_0 can be written in the form

$$
\begin{aligned}
z &= \int_0^{t_1} w_1 \, dt + \int_{t_1}^{t_2} w_2 \, dt + \cdots + \int_{t_{N-1}}^{T} w_N \, dt \\
&= \sum_{r=1}^{N} w_r(t_{r-1} - t_r),
\end{aligned}
\tag{6.1}
$$

and similarly for the other coordinates. The times $t_0 = 0$ and $t_N = T$ need not correspond to collisions. The average value of z vanishes since the average value $\langle w_r \rangle = 0$ by assumption in each term of (6.1). A nontrivial result is obtained by considering $\langle z^2 \rangle$, giving

$$
\langle z^2 \rangle = \sum_{s=1}^{N} \sum_{r=1}^{N} \langle w_s w_r (t_{s-1} - t_s)(t_{r-1} - t_r) \rangle.
$$

Considering in particular a short time interval T within which collisions can be ignored and therefore $z = \int_0^T w_1 \, dt = T w_1$, we find

$$\langle z^2 \rangle = \langle w_1^2 \rangle T^2. \tag{6.2}$$

This short time limit, (6.2), is determined by purely inertial effects, independent of the collision frequency.

Consider now the case in which the particle has undergone many collisions. Since velocities in different time intervals are statistically independent and also independent of the actual time interval, we have

$$\langle z^2 \rangle = \sum_{s=1}^{N} \sum_{r=1}^{N} \langle w_s w_r \rangle \langle (t_{s-1} - t_s)(t_{r-1} - t_r) \rangle$$

$$= \langle w^2 \rangle \sum_{s=1}^{N} \langle (t_{s-1} - t_s)^2 \rangle$$

$$= 2 \frac{\kappa T}{m} \frac{N}{v^2}.$$

For a light particle we assumed $\langle w_s w_r \rangle = \langle w_s^2 \rangle \delta_{r,s} = \langle w^2 \rangle \delta_{r,s}$ in terms of the Kronecker delta, with the arguments used in Section 3.1. The particle's velocity after a collision is assumed to be entirely uncorrelated to the one it had before the collision. (This assumption will not hold for *heavy* particles moving in a background of light ones.) We used $\langle (t_{s-1} - t_s)^2 \rangle = 2/v^2$ for all s. With the assumption of the particle being in thermal equilibrium after one collision, we also used $\langle w^2 \rangle = \kappa T/m$. The number of collisions, N, in the series is statistically varying over the realizations and, performing the corresponding averaging, we have $\langle N \rangle = Tv$. Consequently

$$\langle z^2 \rangle = 2 \frac{\kappa T}{m} \frac{T}{v}. \tag{6.3}$$

The limit of large T is often called the diffusion limit with reference to the solutions of the standard diffusion equation

$$\frac{\partial P(z, t)}{\partial t} = D \frac{\partial^2 P(z, t)}{\partial z^2}, \tag{6.4}$$

for the probability density $P(z, t)$ for the particle's displacement z, where D is a constant diffusion coefficient. Recall that the dimension of $P(z, t)$ is length^{-1}. The solution to (6.4) is $P(z, t) = (4\pi Dt)^{-1/2} \exp[-z^2/(4Dt)]$, for the case in which the particle is known with certainty to be at $z = 0$ at $t = 0$, i.e. $P(t = 0, z) = \delta(z)$. Then $\langle z^2 \rangle \equiv \int_{-\infty}^{\infty} z^2 P(t, z) \, dz = 2Dt$ and this is analogous to (6.3) if we define

$$D = \frac{\langle w^2 \rangle}{v} = \frac{\kappa T}{mv}. \tag{6.5}$$

Note that it is the probability density, not the particle density, which enters (6.4). To introduce a particle density, assume that a large number of particles, N, is released simultaneously at the origin $z = 0$ at $t = 0$, and it is assumed (rather optimistically) that they diffuse independently. In the present one-dimensional model the particle density is then approximately $NP(z, t)$.

Let now an externally imposed, deterministic, dc electric field be acting on the selected electron, just like in Section 3.4. The average particle velocity becomes

$$\langle w(t) \rangle = \frac{eE}{mv}. \tag{6.6}$$

The particle mobility, μ, is defined as the ratio of the average velocity and the imposed electric field, giving

$$\mu = \frac{e}{mv}. \tag{6.7}$$

Evidently, from (6.5) and (6.7) we have

$$\mu = \frac{e}{\kappa T}D, \tag{6.8}$$

which is often called the Einstein relation. In spite of its apparent simplicity it demonstrates an important consequence of the fluctuation–dissipation theorem.

> Diffusion in thermal equilibrium is a random motion mediated through collisions, i.e. microscopic processes. Mobility is a macroscopic quantity, but it has its origin in the very same collisional interactions, the two phenomena being related through the Einstein relation. The fluctuating motion of the light particle is not *caused* by thermal motion; it *is* thermal motion.

The diffusion coefficient D in (6.5) is inversely proportional to mass; this feature can be used for mass separation. Recall, though, that the present results were derived under the assumption that we are dealing with diffusion of a light particle in thermal equilibrium with a gas of heavy particles. Violation of this assumption can have nontrivial consequences.

6.1.1.1 Transition to thermal equilibrium

Assume that we know with certainty that the initial velocity of a particle is w_0. The corresponding series is no longer time stationary; the time $t = 0$ is distinguished from all other times! The average value is now w_0 for $t > 0$ provided that no collision has occurred. After a collision, on the other hand, the velocity is completely random, by assumption, and the average velocity is zero. With the present collision model we thus have

$$\langle w \rangle = w_0 e^{-vt},$$

where e^{-vt} is the probability of there being no collisions in the time interval $\{0; t\}$. From the present time record a new one can be formed by taking the square of $w(t)$ at each time. The ensemble average is w_0^2 provided that no collision has occurred and it is $\kappa T/m$ after one collision or more. At a time t the statistical average of w^2 is therefore

$$\langle w^2 \rangle = w_0^2 e^{-vt} + (1 - e^{-vt})\kappa T/m,$$

where $1 - \exp(-vt)$ is the probability of there being one or more collisions in the time interval $\{0; t\}$. These results demonstrate that thermal equilibrium is reached on a timescale given by v^{-1}, which is after all precisely what was built into the present simple collisional model.

6.1.2   Diffusion of heavy particles

Considering the random collisions of light particles with a heavy gas, it could be justified that, *after* a collision, the velocity component of the light particle along any direction in space could be considered statistically independent of the value it had *before* the collision. For the opposite case, in which a heavy particle, a dust grain for instance, is influenced by the random impacts of light particles, its velocity will be changed but little by an impact due to inertial effects that make this problem different than the one considered previously. The problem was apparently first analyzed by Strutt (who later became Baron Rayleigh) (1902) and heavy particles in such an environment are often called *Rayleigh particles*. The one-dimensional version of the problem is then called a Rayleigh *piston*. As a simple illustration we consider this one-dimensional problem, assuming that we have elastic collisions. For two particles with masses M and m and initial velocities w_1 and u_1, the velocities are $w_2 = (Mw_1 - mw_1 + 2mu_1)/(m + M)$ and $u_2 = (mu_1 - Mu_1 + 2Mw_1)/(m + M)$ after a collision. For the velocity of the heavy particle after the jth collision we have

$$w_j = w_{j-1}\left(1 - 2\frac{m}{M + m}\right) + 2\frac{m}{M + m}u_{j-1}. \tag{6.9}$$

Measuring velocities in the rest frame of the light gas, where $\langle u_j \rangle = 0$, we have

$$\langle w_j \rangle = w_0\left(1 - 2\frac{m}{M + m}\right)^j \to 0 \qquad \text{for } j \to \infty, \tag{6.10}$$

v_0 being the initial, deterministic, velocity of the heavy particle. Its *average* velocity thus slows down to zero after essentially approximately M/m collisions. Note that nothing is said about the *time* it takes; $t_j = \sum_\ell^j \Delta_\ell t$ and the time $\Delta_\ell t$ between the collisions of numbers ℓ and $\ell + 1$ depends on the relative velocity of the particles, in general.

Relaxation of energy can be analyzed similarly by squaring and averaging (6.9), giving

$$\langle w_j^2 \rangle = \langle w_{j-1}^2 \rangle\left(1 - 2\frac{m}{M + m}\right)^2 + \langle u_{j-1}^2 \rangle\frac{4m^2}{(M + m)^2}, \tag{6.11}$$

using that u_j is statistically independent of the actual value of w_j and also that $\langle u_j u_i \rangle = 0$ for $i \neq j$. By iterating (6.11) we find that

$$\langle w_{j+1}^2 \rangle = \langle w_0^2 \rangle\left(1 - 2\frac{m}{M + m}\right)^{2j}$$
$$+ \langle u^2 \rangle\frac{4m^2}{(M + m)^2}\left[\left(\frac{M - m}{M + m}\right)^{2j} + \left(\frac{M - m}{M + m}\right)^{2(j-1)} + \cdots + 1\right], \tag{6.12}$$

since $\langle u_j^2 \rangle = \langle u^2 \rangle$ for all j. Using $(1 - x^2)^{-1} = 1 + x^2 \cdots + x^{2n} \cdots$, we find that $M\langle w^2 \rangle \to m\langle u^2 \rangle$ for $j \to \infty$, as expected.

One problem deserves to be mentioned. It is throughout being assumed that collisions can be identified one by one, and that the particles are influenced by one collision at a time. This is perfectly feasible when the duration of a collision can be taken to be negligible. In reality this approximation can of course not be exact. Any collisional interaction takes *some* time. If the heavy particle is macroscopic, such as a dust grain, it has a large area, $\mathcal{A}$. The number of particles hitting it per unit time is proportional to $\mu \mathcal{A}\langle u^2 \rangle^{1/2}$, where μ is the particle density of the medium surrounding the heavy particle. During the bombardment by the many particles in this surrounding medium, be it gas or liquid, it will in general be necessary to consider the simultaneous impact of several particles when one allows for the finite duration of the collision, when the density μ is large. The analysis to be presented in the following avoids this question entirely.

6.2 Displacement of macroscopic particles

The effects discussed in the previous section can not be observed experimentally; it is not possible to follow the motion of one individual electron or ion. Readily observable small, light particles like pollen and dust will, however, exhibit a somewhat similar motion when they are suspended in a suitable liquid. Typical sizes of pollen grains are 50 μm i.e. they are large enough to be discriminated even with relatively simple microscopes, but light enough to exhibit observable thermal motion. Observations of this motion were reported as early as 1785, by the Dutch physician Jan Ingenhausz, who examined powdered charcoal on an alcohol surface (Klafter *et al.* 1996). The phenomenon was, however, named later after a biologist, Robert Brown (1827, 1828a, 1828b), who gave an extensive report of his detailed investigations of the motion of tiny pollen grains in water observed by using a microscope. He was eventually carried away by his studies, investigated all kinds of materials (including, apparently, dust from a fragment of the Sphinx), and found that this sort of motion occurred for every kind of substance he examined. Brown's own conclusion was seemingly that he had encountered a manifestation of the existence of some elementary form of life in all organic and inorganic matter. For the history of Brownian motion, see for instance the interesting and informative notes by Fürth in Einstein's collected papers on Brownian motion (Einstein, 1956) and also MacDonald (1962). It took a rather long time before the present-day explanation of the phenomenon was accepted, although it had been formulated by several authors as early as the end of the nineteenth century (MacDonald, 1962). Systematic research made it evident that the explanation was to be sought in the collisions with molecules of the surrounding fluid that were in thermal motion. In 1906 Einstein and Smoluchowski both derived essentially the same formula from considerations of kinetic theory (although by different methods, see for instance Lee *et al.* (1973)). Ironically, it is now being argued (Deutsch, 1992) that what Brown observed was *not* what we today call Brownian motion!

6.2.1 Langevin's analysis

Shortly after Einstein's work had been published, Langevin presented in 1908 a different approach that, according to him, was 'infinitely simpler.' His analysis can be summarized as follows. The force on a particle was assumed to be composed of an average and a fluctuating part, $F(t)$ with $\langle F(t) \rangle = 0$, giving

$$M \frac{d^2 x}{dt^2} = -6\pi\eta a \frac{dx}{dt} + F(t), \tag{6.13}$$

where Langevin assumed that the coefficient to the average value $-d\langle x\rangle/dt$ could be identified with the viscous drag given by the same formula as that in hydrodynamics, i.e. $-6\pi\eta av$, where η is the dynamic viscosity of the liquid surrounding the particle, and a is the diameter of the particle, which is assumed to be spherical. The dimension of η is $[\eta] = \text{mass}/(\text{length} \times \text{time})$. On multiplying by x, equation (6.13) can be written as

$$\frac{1}{2}M\frac{d^2x^2}{dt^2} - M\left(\frac{dx}{dt}\right)^2 = -3\pi\eta a\frac{dx^2}{dt} + xF(t). \tag{6.14}$$

As the initial condition it is assumed that the particle is released at the origin. Now, we take the average value of each term in (6.14). By physical reasoning we argue as follows. $F(t)$ is caused by the thermal agitation of the medium surrounding the particle, i.e. an irregular bombardment of molecules. Owing to its large mass, the observed particle changes its velocity but little for each impact. For time intervals of relevance for the movement of the heavy particle we may assume $F(t)$ to be statistically independent of x and, since $\langle F(t)\rangle = 0$ by assumption, it can be argued that $\langle xF(t)\rangle \approx 0$. It is assumed that the initial position of a particle (a pollen or dust grain, or similar) is known with certainty, implying that $\langle x\rangle = 0$ and also $\langle x^2\rangle = 0$ at time $t = 0$. The initial *velocity* of the particle is, on the other hand, unknown. Since the particle is in thermal equilibrium with the surrounding medium, we know that $\langle(dx/dt)^2\rangle \equiv \langle u^2\rangle = \kappa T/M$ for this system in one spatial dimension, which has been used explicitly in (6.15). Consequently

$$\frac{1}{2}M\frac{d^2\langle x^2\rangle}{dt^2} + 3\pi\eta a\frac{d\langle x^2\rangle}{dt} = \kappa T. \tag{6.15}$$

Equation (6.15) has the solution

$$\frac{d\langle x^2\rangle}{dt} = \frac{\kappa T}{3\pi\eta a} + Ce^{-6\pi\eta at/M}, \tag{6.16}$$

where $C = -\kappa T/(3\pi\eta a)$ is a constant determined by the initial conditions. For large times the last term is negligible (in reality for $t > 10^{-8}$ s or so) and Einstein's result

$$\langle x^2\rangle = 2Dt \tag{6.17}$$

is obtained with

$$D = \frac{\kappa T}{6\pi\eta a}. \tag{6.18}$$

It is important to note that the diffusion coefficient D here is independent of the mass M of the macroscopic particle; it enters only in the short transient time for which the second term in (6.16) is important. This particular result was very surprising when it was first obtained. Nevertheless, this prediction of the theory was later show by Perrin (1916) and coworkers to be satisfied to high precision for the substances they investigated. Recall that, for the case of a *light* particle (an electron in Section 6.1.1) moving in a gas of heavy molecules, the diffusion coefficient depended explicitly on the mass. The region of validity of the Langevin analysis was discussed by Lorentz, see also Section 6.2.3, and it was argued that (6.18) has a limited applicability.

For the large particles considered in this section, the diffusion coefficient D in (6.18) depends explicitly on the radius of the particle; this can be used to determine the size of small particles suspended in a fluid by measuring the diffusion coefficient. The method works well for spherical objects; for elongated or irregular ones an average scale size is obtained. The method has been used for measuring the size of viruses, for instance. The analysis predicts that $\langle x^2 \rangle$ is proportional to T. Since, however, viscosity usually decreases rapidly with increasing temperature, the thermal variation of the diffusion coefficient is not easy to verify directly.

The last term in (6.16) can be traced back to the inertia term $M\, d^2x/dt^2$ in (6.13). Since this last term is ignored, we find that Brownian motion is described as *inertialess*, meaning that a particle is supposed to be able to change its direction of motion abruptly. When one is observing Brownian motion with a microscope, this is actually how it appears; the tiny rounding off of the corners of their trajectories due to the last term in (6.16) is hardly visible.

At first sight it seems surprising that $F(t)$ has disappeared completely from the results; apparently it might have been put to zero from the outset. However, it appears indirectly via the identification $M\langle u^2 \rangle = \kappa T$. A random force is necessary in order to mediate the relaxation of the fluctuating velocities to the universal form given by thermodynamic arguments.

A mobility can be defined as the ratio of an average velocity and a deterministic external force. (For historical reasons, the convention is here slightly different than the one used in Section 6.1.) From (6.13) we readily find a velocity $u = F/(6\pi\eta a)$, where F can denote a force due to gravity Mg, for instance. The mobility is thus $\mu = 1/(6\pi\eta a)$ and the ratio $D/\mu = \kappa T$, i.e. the Einstein relation is again recovered.

Langevin's equation can be assumed to have been the first example of a stochastic differential equation to be analyzed. By a stochastic differential is meant one in which a random term appears explicitly and thus it has random functions as solutions.

6.2.2   Brownian motion with a central force

The analysis of the foregoing section can be made somewhat more general by including a central force, giving a Langevin-type equation

$$M\left(\frac{d^2}{dt^2}x + \omega_0^2 x + \gamma \frac{d}{dt}x\right) = F(t), \tag{6.19}$$

where a friction coefficient γ has been introduced. This equation is capable of describing a variety of phenomena, e.g., it is formally equivalent to $L\ddot{q} + R\dot{q} + q/C = e(t)$ with $\omega_0^2 = 1/(LC)$ describing a parallel connection of a resistance R, self-inductance L, and capacitance C while $e(t)$ is a fluctuating voltage and q, as before, is the charge on the condenser. Equation (6.19) might just as well describe the motion of a mechanical system, such as a simple pendulum. For $F(t)$ it is again assumed that $\langle F(t) \rangle = 0$ and $\langle F(t_1)F(t_2) \rangle = R(t_1 - t_2)$. With $x = x_0$ and $u \equiv dx/dt = u_0$ at $t = 0$, one finds by integration, following Uhlenbeck and Ornstein (1930), that

$$u = -\frac{2\omega_0 x_0 + \gamma u_0}{2\omega_1} e^{-\gamma t/2} \sin(\omega_1 t) + u_0 e^{-\gamma t/2} \cos(\omega_1 t)$$
$$+ \frac{1}{\omega_1}\int_0^t F(\tau)e^{\gamma(\tau - t)/2}\left(\omega_1 \cos[\omega_1(t - \tau)] - \frac{\gamma}{2}\sin[\omega_1(t - \tau)]\right)d\tau, \tag{6.20}$$

$$x = \frac{2u_0 + \gamma x_0}{2\omega_1} e^{-\gamma t/2} \sin(\omega_1 t) + x_0 e^{-\gamma t/2} \cos(\omega_1 t)$$

$$+ \frac{1}{\omega_1} \int_0^t F(\tau) e^{\gamma(\tau - t)/2} \sin[\omega_1(t - \tau)] \, d\tau, \tag{6.21}$$

where $\omega_1^2 = \omega_0^2 - \gamma^2/4$. Note that there is no *a priori* reason that x_0 should correspond to the equilibrium position. From the foregoing expressions one obtains immediately

$$\langle x \rangle = \frac{2u_0 + \gamma x_0}{2\omega_1} e^{-\gamma t/2} \sin(\omega_1 t) + x_0 e^{-\gamma t/2} \cos(\omega_1 t), \tag{6.22}$$

demonstrating that the average displacement of the harmonically bound particle goes to zero with the time constant $\gamma/2$. In an individual realization of the ensemble, this displacement is of course nonzero in general, as is evidenced by the standard deviation, which can be obtained as

$$\langle x^2 \rangle = \left(\frac{2u_0 + \gamma x_0}{2\omega_1} e^{-\gamma t/2} \sin(\omega_1 t) + x_0 e^{-\gamma t/2} \cos(\omega_1 t) \right)^2$$

$$+ \frac{\tau_1}{2\omega_1^2 \gamma} (1 - e^{-\gamma t})$$

$$- \frac{\tau_2}{8\omega_0^2 \omega_1^2} \{ \gamma[1 - e^{-\gamma t} \cos(2\omega_1 t)] + 2\omega_1 e^{-\gamma t} \sin(2\omega_1 t) \}, \tag{6.23}$$

introducing $\tau_1 = \int_{-\infty}^{\infty} R(\tau) \cos(\omega_1 \tau) \, d\tau$ and $\tau_2 = \int_{-\infty}^{\infty} R(\tau) \, d\tau$. In agreement with the foregoing arguments, we expect that $R(\tau)$ is sharply peaked around zero, which means that $\tau_1 \approx \tau_2$. On calculating the velocity of the particle and using that for large times, thermal equilibrium is reached, i.e. using $\langle (dx/dt)^2 \rangle = \kappa T/M$, it is demonstrated that $\tau_1 = \tau_2 = 2\gamma\kappa T/M$, and by insertion the problem is solved. This result would have been obtained from the outset by assuming that $\langle F(t_1)F(t_2) \rangle \approx 2D\delta(t_1 - t_2)$ with $D = \kappa T/\gamma$ as before. In particular, the mean-square displacement of the particle becomes $\langle x^2 \rangle = \kappa T/(M\omega_0^2)$ for large times when the particle has reached equilibrium with the surroundings. The particle is thus confined to a finite region, in contrast to the free Brownian motion discussed before, but, if the restoring force is made weaker, i.e. $\omega_0 \to 0$, the mean-square displacement increases.

- **Exercise:** Obtain the expressions for $\langle u \rangle$, $\langle u^2 \rangle$, and $\langle ux \rangle$ from (6.20) and (6.21).

If we now assume that the particle is taken from actual realizations of the ensemble, it is evident that the initial velocity has the same statistics as that caused by the thermal fluctuations. The problem has a natural reduction obtained by taking an average over all realizations of u_0, subject to this condition. Physically, a sub-ensemble is selected by choosing realizations subject only to the condition on the initial position of the particle, and arbitrary, randomly distributed, initial velocities. With the present assumption we now have $\langle u^2 \rangle = \kappa T/M$ at all times.

Other averages than those discussed previously can be calculated. It may for instance be instructive to consider also

$$\langle xu \rangle = \frac{1}{\omega_1 \omega_2} \left(\frac{\kappa T}{M\omega_0^2} - x_0^2 \right) \left(\cos(\omega_1 t) + \frac{\gamma}{2\omega_1} \sin(\omega_1 t) \right) e^{-\gamma t} \sin(\omega_1 t), \tag{6.24}$$

with the initial velocities statistically distributed as discussed before. The correlation between position and velocity thus starts out being zero, then oscillates and eventually vanishes.

- **Exercise:** Discuss the damped harmonic oscillator subject to thermal agitation described by the foregoing analysis for the overdamped, critically damped, and underdamped cases.

6.2.3   Limitations of the Langevin equation

It was noted already by Lorentz (1921) that the Langevin equation in its simplest form (6.13) or (6.19) can be applied within certain limits only. His arguments were essentially based on a dimensional analysis that is quite instructive. Introduce first a time scale τ_B for the relaxation time of the velocity correlation function for the Brownian particle. In terms of the viscosity η we have as an order of magnitude $\tau_B \simeq M/(\eta R)$, where R is the typical scale size of this particle. Another timescale can be introduced, namely one characterizing the relaxation of a change in velocity of the fluid as a response to the motion of the particle. The time it takes a hydrodynamic perturbation to relax over a length scale R is $\tau_H \sim R^2/\nu$, where ν is here the kinematic viscosity, $\eta \approx \nu \rho_F$, where ρ_F is the mass density of the fluid. Diffusive relaxation was assumed for the hydrodynamic velocity; for the present condition this is the most reasonable hypothesis. In order to have a well-defined fluid velocity that is the same all over the surface of the Brownian particle, as is implicitly assumed in (6.13), it is evidently necessary that $\tau_H \ll \tau_B$, i.e. $R^2 \ll \nu M/(\eta R)$, or

$$\rho_F \ll M/R^3 \simeq \rho_B, \tag{6.25}$$

where ρ_B is the mass density of the Brownian particle, implying that the Langevin equation applies for particles heavier than the fluid. The inequality (6.25) is not always satisfied and the analysis has been extended to more general cases, e.g. by Hauge and Martin-Löf (1973).

It might be noted that the inequality (6.25) need not imply that the heavier particles fall to the bottom of the fluid, so to speak. The particles participate in the thermal motion while being influenced by gravity as well as the lift of the surrounding fluid, and will obtain a height distribution with a scale length $\mathcal{L} = [\kappa T/(Mg)]/(1 - \rho_F/\rho_B)$, see for instance Chandrasekhar (1943) and also Section 11.1.1. Provided that this scale length is macroscopic, it is meaningful to expect the thermal fluctuations to be well described by the Langevin equation. Of course, with tiny lead spheres in water this scale length will be comparable to the diameter of the spheres and the problem becomes meaningless.

6.3   Galvanometer fluctuations

A D'Arsonval galvanometer, see Fig. 6.1, is a sensitive instrument for measuring very small currents. The moving system of such a galvanometer consists basically of a coil of wire and a mirror, suspended by a fine fiber and capable of rotation about a vertical axis. The mirror is deflecting a light beam and the movements of the light spot can be used for accurate measurements of the changes in currents through the coil which give rise to the movements of the mirror. In a way the galvanometer can be understood as an ammeter in which the light beam plays the role of a very long, weightless, arm. It is evidently essential for the proper functioning

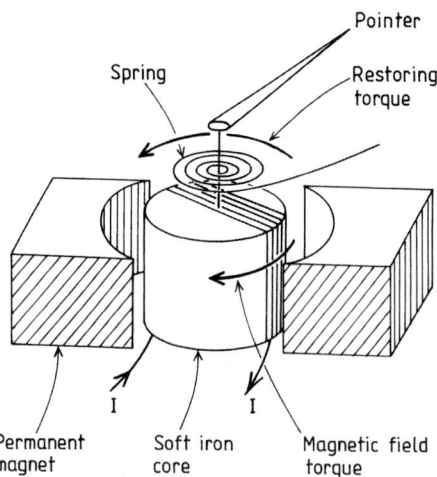

Spring

Pointer

Restoring
torque

I

I

Permanent
magnet

Soft iron
core

Magnetic field
torque

Figure 6.1 A simplified presentation of a D'Arsonval galvanometer. Owing to the cylindrical soft iron core, the magnetic field in the gap is approximately radial, with constant intensity at a fixed radial position. The role of the pointer, drawn here for simplicity, is, in sensitive galvanometers, taken by a small mirror that deflects a light beam.

of the instrument that the coil, mirror and supports are made of light material in order to ensure that a rapid response occurs. Thermal motions can become of importance and will impose a lower limit on the displacement of the light spot which can be discriminated with accuracy. Random collisions of air molecules with the suspended system produce torques that are not equal and opposite at all instants. The result is that the angular position of the system is continuously fluctuating, and the system exhibits an unsteady zero, another example of Brownian motion.

The moving parts of the galvanometer can be considered as a very large Brownian particle, which participates in the thermal motion but is subject also to a central restoring force. In this case, however, it can safely be argued that the inequality (6.25) is satisfied. Assuming that the friction in the mounts can be neglected, we have air damping of the galvanometer corresponding to the viscous friction force on a particle in suspension. This problem was analyzed by, e.g., Uhlenbeck and Goudsmit (1929) using methods proposed in a different context by Schottky (1918). In the present outline we follow Feynman *et al.* (1963) and Lee *et al.* (1973). It serves as an example of the interplay between a mechanical and an electronic instrument, in particular for the coupling between electric and mechanical thermal fluctuations.

The dynamics of the galvanometer are most conveniently described in terms of its angular deflection θ and its angular velocity $\omega \equiv d\theta/dt$. With K the torque constant, which has to be small for a sensitive instrument, while I is the moment of inertia, we have

$$I\frac{d^2}{dt^2}\theta = -K\theta, \tag{6.26}$$

see also (4.1). The natural period of oscillation for the freely swinging galvanometer is $\tau_0 = 2\pi\sqrt{I/K}$. The expression for the root-mean-square (rms) displacement or the angular velocity can immediately be written down, with reference to the law of equipartition, since both the elastic potential energy $\frac{1}{2}K\theta^2$ and the rotational kinetic energy $\frac{1}{2}I\omega^2$ are proportional to the square of their respective coordinates, and the mean energy associated with each is $\frac{1}{2}\kappa T$. Thus we have $\frac{1}{2}I\langle\omega^2\rangle = \frac{1}{2}\kappa T$ and $\frac{1}{2}K\langle\theta^2\rangle = \frac{1}{2}\kappa T$, implying for instance that the rms angular displacement is

$$\theta_{\text{rms}} \equiv \sqrt{\langle\theta^2\rangle} = \sqrt{\kappa T/K}. \tag{6.27}$$

Note that, the smaller the torque constant K and the more sensitive the galvanometer, the larger are the statistical fluctuations of the zero point.

- **Exercise:** The torque constant K for a fine quartz fiber is of the order of 10^{-6} dyne cm rad^{-1} or 10^{-13} N m rad^{-1}. Assume that the light source and scale are at a distance of 1 m from the mirror. Calculate the rms fluctuation of the light spot when the system is at a temperature of 300 K.

To detect a current with a galvanometer, one must be reasonably certain that an observed deflection is not caused just by the impact of an atom or a molecule from the surrounding neutral gas. First, a general expression for the minimum detectable current is obtained in terms of the galvanometer constants. Let the coil have N turns of area $\mathcal{A}$, and swing in a homogeneous magnetic field of flux density B, see also Fig. 6.1. On evaluating the $i \times B$ force on the four sides of the rectangular coil, we find nonvanishing contributions only from the two in the gap between the soft iron core and the permanent magnet. With the plane of the coil parallel to the magnetic field, the deflecting torque is readily obtained as $M_d = Ni\mathcal{A}B = iN\Phi$, in terms of the current i, where we have replaced $\mathcal{A}B$ by Φ, this being the total magnetic flux intercepted by the coil when it is normal to the field. The restoring torque at an angular deflection θ is $M_r = K\theta$. In the equilibrium position these two torques are equal, $M_d = M_r$. Consequently

$$\theta = \frac{N\Phi}{K}i. \tag{6.28}$$

We can introduce a quantity

$$R_c = \frac{N^2\Phi^2\tau_0}{4\pi I}, \tag{6.29}$$

which has the dimension of resistance. It can be shown (Lee et al., 1973) that the critical external damping resistance is given by R_c in (6.29), and we shall assume that the instrument is shunted by this resistance.

Assuming that the random deflections follow a Gaussian distribution, the fractional number of a deflection greater than any pre-assigned value can be calculated. It is readily shown that on average only one deflection out of 3,000 is greater than four times the rms deflection, so we take $4\theta_{\text{rms}}$ as a limit and can then be reasonably certain that any observed deflection greater than this is not due to chance. To find the minimum current i_{min} that can be detected with adequate certainty, we therefore set θ in (6.28) equal to $4\theta_{\text{rms}}$:

$$\frac{N\Phi}{K}i_{\text{min}} = 4\theta_{\text{rms}}.$$

Using (6.27), (6.29), and the definition of τ_0, this becomes

$$i_{\text{min}} = 4\sqrt{\pi\kappa T/(R_c\tau_0)}. \tag{6.30}$$

- **Exercise:** Find i_{min} for a typical case in which $\tau_0 = 15$ s and $R_c = 100\,\Omega$ at room temperature 300 K.

Thus, although the statistical fluctuations of θ increase inversely with $\sqrt{K}$, see (6.27), the minimum detectable current can be decreased by using a long-period instrument, since the natural period τ_0 is inversely proportional to $\sqrt{K}$. The limit is set by the mechanical stability of an extremely fine fiber and the patience of the observer.

The galvanometer has two contributions to its damping, namely its resistance R in the circuit and the viscosity of the surrounding gas. Both these mechanisms are due to thermal agitations; the first is due to those in the conductors for the detected current and the second is due to the thermal motion in the gas. Both give rise at the same time also to the galvanometer fluctuations. The Brownian fluctuations of a galvanometer can not be removed by suspending the mirror in a vacuum, thereby eliminating the fluctuating torques produced by molecular collisions. Apart from the fact that, even at the low pressure produced by the best pumping systems, there still remains a tremendous number of molecules per unit volume, we see that such a procedure would not produce the desired result, not even in principle. Consider first the effect of reducing the pressure. If the galvanometer system is in thermal equilibrium with a heat reservoir at temperature T, it must have a mean rotational kinetic energy of $\frac{1}{2}\kappa T$. If the system, which initially is at rest, receives an angular impulse from a molecular collision, it starts to oscillate at its natural frequency and continues to do so until it is disturbed by the next impulse. When the density of the surrounding gas is relatively high, these impulses come in such rapid succession that the natural oscillation can not be detected: the system is heavily overdamped. As the pressure is lowered, the intervals between collisions become longer, and the motion tends more and more toward that of an isolated system: the system is underdamped. The rms amplitude, however, remains the same. Experimentally obtained curves illustrating the time-varying signals were published by Gerlach (1927), see also Lee *et al.* (1973) or Pathria (1996). The curves were registered at different pressures, and are quite different in appearance, but have the same mean-square deviation, see for instance Fig. 6.2.

Even if the colliding *molecules* could be completely eliminated by pumping down the system, there would still be a flow of energy between the system and its surroundings by radiation. This radiant energy is carried by light quanta or photons, which have energy and momentum, and whose emission and absorption are statistical phenomena. The emission or absorption of a photon produces a change in angular momentum of the system, and the rms energy of oscillation continues to be $\frac{1}{2}\kappa T$. Reducing the temperature of the surroundings reduces the amplitude of the fluctuations, which approaches zero as the temperature

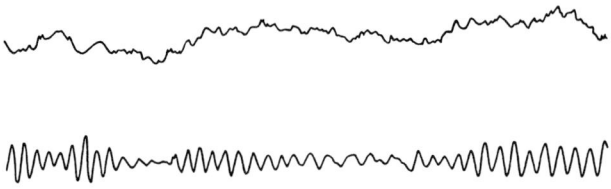

Figure 6.2 Two oscilloscope traces of the thermal oscillations of a galvanometer-mirror system suspended in air. The upper trace was obtained at atmospheric pressure, the lower one at a pressure of 10^{-4} mm of mercury.

approaches 0 K. The oscillations of a *freely swinging* galvanometer can be reduced in this way. However, if the instrument is connected to an external circuit at finite temperature, the current through the system is carrying with it its thermal fluctuations of charge from the external circuit. This is the Johnson or Nyquist noise discussed in Chapters 3 and 4. If the temperature of the external circuit is T, the mean-square energy of the galvanometer is still $\frac{1}{2}\kappa T$ even if the effect of the neutral gas surrounding the galvanometer could be removed completely. We model the galvanometer coil here as a single wire in the form of a closed loop, having a resistance R and self-inductance L. Capacitive effects are neglected. Assume that the random motion of electrons in the wire is equivalent to a fluctuating electromotance $\mathcal{E}$ in the circuit. Then, at any instant,

$$L\frac{d^2q}{dt^2} + R\frac{dq}{dt} = \mathcal{E},$$

where the current is $i \equiv dq/dt$, in terms of q being the net charge crossing any section of the circuit. This equation has exactly the same form as that for the displacement of a particle in Brownian motion, and we can follow the procedure used for solving Langevin's equation (6.13). The equation is multiplied throughout by q, and the mean magnetic energy $\frac{1}{2}L\langle i^2\rangle$ set equal to $\frac{1}{2}\kappa T$. We assume again that, because of the random nature of $\mathcal{E}$, the mean value $\langle \mathcal{E}q \rangle$ vanishes. The final result is

$$\langle q^2 \rangle = \frac{2\kappa T}{R}\tau, \tag{6.31}$$

where τ is the duration of the time interval or time record in question. The physical significance of $\langle q^2 \rangle$ must be borne in mind; i.e., for each of a number of equal time intervals of duration τ, we find the net charge q that has crossed the section during this interval, then we square these, and finally take their average value. Equation (6.31) states that this average is proportional to the duration of the time interval and to the temperature, while it is inversely proportional to the resistance. The self-inductance L drops out, as did the mass M in Brownian motion, showing that the shape of the circuit has no effect (except that the time interval τ must be much larger than the time constant L/R).

The current associated with any one of the time intervals is, as already stated, $i = q/\tau$, where q is the net charge crossing during that interval. These currents are of course in general different for different time intervals. The square of the current associated with any one interval is $i^2 = q^2/\tau^2$, and the average of these over an ensemble of intervals, all with the same duration, is $\langle i^2 \rangle = \langle q^2 \rangle/\tau^2$. Using (6.31), we have

$$\sqrt{\langle i^2 \rangle} = \sqrt{2\kappa T/(R\tau)}. \tag{6.32}$$

Except for the numerical factor, the right-hand side of this equation has exactly the same form as that of (6.30) if we let $\tau = \tau_0$, the natural period of the galvanometer, and $R = R_c$. That is, the natural fluctuations of the electrons in the external circuit set the same lower limit on the current that can be detected by a galvanometer as do the oscillations of the galvanometer's mirror, provided that the circuit and the mirror are at the same temperature (Lee *et al.*, 1973).

6.3.1   The human ear

The human eardrum, together with the bony structure coupling it to the inner ear, can be considered as a mass–spring system, in some ways reminiscent of the galvanometer. This system will also fluctuate with thermal motion. With a mass M and stiffness constant K, the rms amplitude of the thermal motion is $\sqrt{\kappa T/K}$, this quantity being of the order of 10^{-8} cm for the present problem, at a characteristic frequency of $[1/(2\pi)]\sqrt{K/M}$. Sounds so faint that they drive the eardrum with an amplitude smaller than this are drowned out by the internal thermal noise. In practice this thermal noise level sets the limit (Morse, 1969) on the sensitivity of the human ear, at least in the frequency range of greatest sensitivity, 1–3 kHz. If the incoming noise level is smaller than this, we hear the thermal fluctuations of our eardrums rather than the outside noise.

6.4   A *perpetuum mobile* of the second kind

The intriguing question of whether it is possible to extract energy from the thermal molecular motion to create a *perpetuum mobile* of the second kind arises. (We met this temptation already when discussing electric networks containing a diode.) At first sight it seems possible to invent a device that will violate the Second Law of Thermodynamics, that is, a gadget that will generate work from a heat reservoir with everything at the same temperature.

To be specific, we assume that we have a box with a gas at a certain temperature, and inside there is an axle with vanes in it, see Fig. 6.3. The vanes and the shaft can be considered as gigantic Brownian particles; because of the bombardment of the vanes by gas molecules, the shaft will oscillate and jiggle, just like the galvanometer discussed in Section 6.3. All that seems necessary is to hook onto the other end of the axle a wheel that can turn only one way, as illustrated in Fig. 6.3. When the shaft tries to jiggle one way, it will not turn, and when it jiggles in the other way, it *will* turn. At least in principle one might tie a weight onto a string wound

Figure 6.3 The ratchet-and-pawl machine. Details of the wheel are shown on the right-hand side.

around the shaft, and make the device produce work. Now, according to Carnot's hypothesis, this is impossible. The fact that no more than a certain amount of work can be extracted on going from one temperature to another is deduced from another axiom, which states that, if everything is at the same temperature, heat can not be converted to work by means of a cyclic process. The apparent paradox is resolved by studying in detail a device that can make the shaft turn in one way only. It will here be discussed for one particularly simple and appealing example based on a ratchet and a pawl that was discussed by Feynman *et al.* (1963).

At first sight, when considering the device shown schematically in Fig. 6.3, it might seem operative. On taking a closer look at the ratchet and pawl, there are, however, several complications. First, the idealized ratchet is as simple as possible, but, even so, there is a pawl, and there must be a spring acting on the pawl to keep it in contact with the ratchet. The pawl must return after coming off a tooth, so the inclusion of a spring is necessary. Another feature of this ratchet and pawl is quite essential. Suppose first that the device was made of perfectly elastic parts. If the pawl is lifted off the end of the tooth and is pushed back by the spring, it will bounce against the wheel and continue to bounce. Then, when another fluctuation came, the wheel could turn the other way, because the tooth could get underneath while the pawl was up. An essential part of the irreversibility of the wheel is therefore a damping mechanism that stops the bouncing. For explicitness, we assume here that the spring is damped, while the friction between the pawl and the ratchet can be ignored. Owing to the damping, the energy that was in the pawl eventually goes into the spring in the form of heat. So, as the wheel turns, the spring will get hotter and hotter.

The pawl and spring, both at some temperature T, also have thermal motion. Consequently, once in a while the pawl has enough energy to overcome the pressure of the spring so as to lift itself above the ratchet. The probability that the pawl lifts itself above a tooth of the ratchet by virtue of its thermal motion is written as $e^{-\epsilon/(\kappa T)}$. During a short time interval, as long as the pawl is up, the wheel is free to turn in both directions, depending on the impacts of the molecules on the vanes. The motion is such that, every once in a while, by accident, the pawl lifts itself up and over a tooth just at the moment when the Brownian motion on the vanes is trying to turn the shaft backward. As the spring gets hotter, this happens more often. The chance that the system can accumulate enough energy to get the pawl over the top of the tooth by bombardments of the vanes is proportional to $e^{-\epsilon/(\kappa T)}$ as before. However, the probability that the pawl will accidentally be up due to its own thermal motion is also proportional to $e^{-\epsilon/(\kappa T)}$ with the same proportionality constant. The number of times that the pawl is up and the wheel can turn backward freely is, on average, equal to the number of times that we have enough energy to turn it forward when the pawl is down. The result is a balance, and the wheel will not go around on average. The ideas outlined here can be pursued in even greater detail, with somewhat surprising results (Feynman *et al.*, 1963).

7   **Random walks**

A very simple, yet illustrative, model for describing diffusion processes is called the *random-walk* model (Cox and Miller, 1965, Feller, 1968). In its simplest version it is given as a discrete grid in one, two, or three dimensions, such that an object is moving by one step at discrete times, the direction of each step, and possibly also its magnitude, being randomly selected. For rather self-evident reasons the process is sometimes called a *drunkard's walk*. As an illustration it is simplest to consider the problem in one spatial dimension, i.e. a line with discrete and equally spaced positions. Assume that an object, say a pawn from a chess game, is placed at the origin and then moved by one step at regular time intervals, such that the direction of the step is decided by the flipping of a coin. The process can evidently be visualized in a presentation like the one in Fig. 7.1 with points in a plane spanned by a displacement axis and a time axis. The problem is formally identical to a classical game of coin tossing; if tails comes up, player A wins, whereas if it is heads, then player B is the lucky one. The equivalent of the displacement axis is then one for cumulative gain (or loss), taking for instance the positive direction to be the one favoring player A. This formulation of the problem has proved appealing to the imagination and is often found in the literature (Feller, 1968).

7.1   Simple random walks

The problem is that of how to obtain for instance the probability of finding the moving object, e.g. a particle, at a position N steps from the origin after S intervals, or, if you like, the probability of the corresponding gain for player A. For definiteness, we introduce the notation of time being measured in multiples of Δ_t, i.e. $t = n\Delta_t$, and displacement measured in units of Δ_x, i.e. a position $X = r\Delta_x$, with n and r integers. A position in space–time is thus given by (r, n). Often (Feller, 1968) the word *epoch* is used to denote points on the time axis, and the word *time* will then refer to an interval or duration. The starting point for a walk will always be $(0, 0)$.

Assume that a walk at time n consists of p steps in the positive direction and q negative steps, $n = p + q$. The partial sum $s_k = \varepsilon_1 + \cdots + \varepsilon_k$ represents the difference between the numbers of steps in the positive and the negative directions at time k, ε_k taking the value $+1$ or -1 depending on the direction of the step with label k. Then $s_k - s_{k-1} = \pm 1$, with $k = 1, 2, \ldots, n$, $s_0 = 0$, and $s_n = p - q$. A path is defined as a polygonal line with vertices at abscissas $0, 1, \ldots, n$ and ordinates $s_0, s_1, \ldots, s_n$, where $s_n = r$. We shall refer to n as the *duration* of a path. There are 2^n paths of duration n. If, among all ε_k, p are positive and q are negative, then $n = p + q$ and $r = p - q$. A path from $(0, 0)$ to (r, n) will exist only if n and r are of this form. In this case (Feller, 1968) the p places for the positive ε_k's can be chosen from the $n = p + q$ available places in

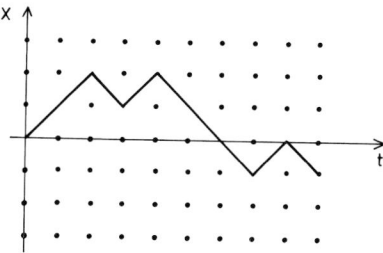

Figure 7.1 An illustration of the space–time variation of a simple random walk. Points on the x-axis (vertical) are labeled r, those on the time axis are labeled by n.

$$N_{r,n} = \binom{p+q}{p} = \binom{n}{\frac{1}{2}(n+r)} \tag{7.1}$$

different ways, expressed in terms of the binomial coefficient, see also Appendix A. We define $N_{r,n} = 0$ whenever n and r are not of the form implied in (7.1), i.e. it is understood that the binomial coefficient is to be interpreted as zero, unless $\frac{1}{2}(n+r)$ is an integer between 0 and n, both inclusive. It takes at least $n = p$ steps to reach a position $r = p$, implying in particular that $N_{r,n} = 0$ if $r > n$. With this convention, there exist exactly $N_{r,n}$ different paths from the origin to a point (r, n). The probability that the position with label r is actually reached at time n is then

$$P_{r,n} = \binom{n}{\frac{1}{2}(n+r)} \frac{1}{2^n}, \tag{7.2}$$

recalling that there are 2^n paths available altogether for a given n, since at each time we have two choices for the direction of a step. Just like for $N_{r,n}$, we have that $P_{r,n} = 0$ for $r > n$.

- **Example:** Since X at time $t = n\Delta_t$ is the sum of n independent variables x_i, with $i = 0, 1, \ldots, n$, where $\langle x_i \rangle = 0$, $\langle x_i x_j \rangle = 0$ for $i \neq j$, and $\langle x_i^2 \rangle = \Delta_x^2$, it is evident that the average is $\langle X \rangle = 0$ and the root-mean-square displacement is $\langle X^2 \rangle^{1/2} = \Delta_x \sqrt{n}$. With time $t = n\Delta_t$ we have $\langle X^2 \rangle = t\Delta_x^2/\Delta_t$ proportional to t, which is a characteristic of diffusion processes.
- **Example:** When asked questions on the edge of, or even beyond, one's competence, it is often tempting to make a wild guess. Alternatively, an answer can be argued out by positing some lengthy and uncertain chain of arguments. Intuitively, it could be expected that the ultimate uncertainty in the answer might be the same, but, as the following arguments attempt to indicate, it need not be so. Enrico Fermi is quoted for illustrating this problem for his students by asking them to estimate the number of piano tuners in Chicago. (This author feels more familiar with conditions in Oslo, so the following discussion refers to that city.) At first sight the answer might be anybody's guess, but it is possible to argue out a fairly accurate answer with even rather modest *a priori* knowledge. The argument consists of several steps. (*i*) The number of inhabitants in Oslo and its near surroundings is probably close to a million, a little less than a quarter of the Norwegian population. (*ii*) In the author's experience, most Norwegian families consist of two parents and two children, sometimes more, sometimes less. (*iii*) The population of the city of Oslo then probably consists of 250,000 families. (*iv*) Again from the author's experience, one out of four,

or one out of five, families possesses a piano, giving 50,000–60,000 pianos in Oslo. (*v*) Depending on its use, it is probably wise to tune a piano once a year, but it is like going to the dentist; one does it probably every second year, or less often, on average. (*vi*) With five working days per week there are approximately 250 working days per year, allowing for vacations etc. We can then estimate the number of pianos being tuned in Oslo per day to be around 100. (*vii*) On looking into a piano you see an immense number of strings, and it probably takes a piano tuner about half a day to tune one, given time for lunch and transport, i.e. two pianos tuned per day. This implies that there should be around 50 piano tuners active in Oslo and its surroundings. On checking the yellow pages in the city's telephone directory (no cheating!), we find 43 piano and organ tuners listed; a result amazingly close to the previous estimate, even though the reasoning consisted of seven more or less uncertain partial arguments. The point is that, although the errors *can* add up, they are most likely to cancel out sometimes, and we have a sort of random walk in errors, which can give a better result than would just an unqualified wild guess.

The question of the return to the origin at time *t* is a special case with $r = 0$ in (7.2). This particular probability is used frequently so it is given a special symbol:

$$U_{2n} = \binom{2n}{n} \frac{1}{2^{2n}} = \frac{(2n)!}{2^{2n}(n!)^2},$$

(7.3)

where it was used that, for this particular case, the argument in (7.2) is necessarily even, and therefore we introduced $2n$ instead of n. The variation of U_{2n} with n is shown in Fig. 7.2. On expressing the binomial coefficient in terms of factorials, use of Stirling's formula $n! \approx \sqrt{2\pi}e^{-n}n^{n+1/2}$ shows that $U_{2n} \approx 1/\sqrt{\pi n}$ for large n.

- **Example:** A simple proof of Stirling's formula can be outlined as follows. First write $\ln n! = \ln 1 + \ln 2 + \cdots + \ln n$. For large n this sum is approximated by the integral $\int_1^n \ln y \, dy$. By partial integration we find $\ln n! \approx \int_1^n \ln y \, dy = n \ln n - n + 1$. By taking the exponential we obtain an expression that, for large n, approximates the relation given before.
- **Exercise:** Demonstrate that the probability that, up to and including time $2n$, *no* return to the origin has occurred is also given by U_{2n} defined in (7.3). (Hint: prove

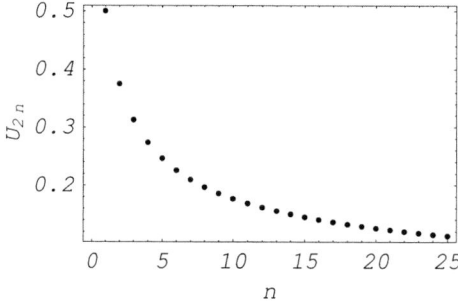

Figure 7.2 The variation of U_{2n} with n given by (7.3).

first that there are exactly $(x/n)N_{x,n}$ paths $(s_1, \ldots, s_n = x)$ from the origin to the point (x, n) such that $s_1 > 0, \ldots, s_n > 0$.)

From (7.3) we can obtain the probability $\alpha_{2k,2n}$ that, up to time t, corresponding to an epoch $2n$, the last visit to the origin occurs at time $2k$. The result is obtained by arguing that U_{2k} is the number of favorable paths divided by the total number of paths up to $(0, 2k)$, the total number of paths being 2^n as before. The next $2n - 2k$ paths can be chosen in $U_{2n-2k}2^{2n-2k}$ different ways. Consequently

$$\alpha_{2k,2n} = U_{2k}U_{2n-2k}, \tag{7.4}$$

for $k = 0, 1, \ldots, n$. Note that the distribution is symmetric in the sense that $\alpha_{2k,2n} = \alpha_{2n-2k,2n}$. By use of Stirling's formula it is demonstrated that $\alpha_{2k,2n} \approx 1/[\pi\sqrt{k(n-k)}]$ with a negligible error, unless k is close to n or 0, which are the two maxima for $\alpha_{2k,2n}$. For $k = 0$ or $2n = 2k$ we have $\alpha_{0,2n} = \alpha_{2n,2n} = U_{2n}$, as given by (7.3).

With some algebra (Feller, 1968) it can be demonstrated that the probability that, in a time interval from 0 to $2n$, a random walk spends $2k$ time units on the positive side and $2n - 2k$ units on the negative side is $\alpha_{2k,2n}$. The time intervals are not necessarily continuous. (The total time spent on the positive side is necessarily an even number of time units.) It is evident that the probability of $2k$ being equal to n, i.e. half the time being spent on one side and the rest on the other, is small, and in particular smaller than the probability of $2k = 2n$ or $k = 0$. These results have rather important implications; they mean that an object executing a random walk will most probably remain on the positive or the negative side of the origin for extended times, or, if you like; in a long coin-tossing game it is quite likely that one of the players will remain practically the whole time on the winning side, and the other on the losing side (Feller, 1968).

One should expect, naively, that, in, for instance, a prolonged coin-tossing game, the observed number of *changes* in lead should increase approximately linearly with the duration of the game; a doubling of the duration should imply that a selected participant leads approximately twice as often as before. This intuitive reasoning is false (Feller, 1968). The number of *changes* of lead in n trials increases as $\sqrt{n}$ for large n; in $100n$ trials one should expect only ten times as many changes of lead as in n trials! As an extension of a detailed investigation of this problem it can be shown that the probability of r changes of sign in lead in n trials decreases with r. This means that, regardless of the number of tosses of a coin, the event that the lead *never* changes is more probable than is any other *preassigned* number of changes.

This result may appear quite in contradiction to expectation, and it can be assumed that it is this failure of intuition which causes many gamblers' ruin. The results of a single realization shown in Fig. 7.3 illustrate the phenomenon described here. Note the long time sequences spent on one side of the zero axis.

- **Example:** The probability that in, say, 20 tosses of a coin the lead never passes from one player to the other is approximately 0.352, see for instance Fig. 7.2. The probability that the lucky player leads 16 times or more is approximately 0.685. On the other hand, the probability that each player leads ten times is only 0.06.
- **Exercise:** Consider a particle undergoing a random walk in the y-direction on a plane such that the upper half ($y > 0$) moves one step to the right and the lower part ($y < 0$) one step to the left in the x-direction, every time the particle makes one step.

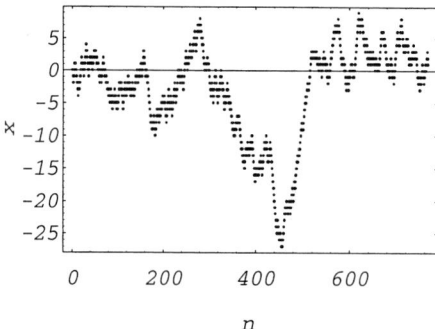

Figure 7.3 A single realization of an extended random walk with 750 steps. Note the long sequences spent on one side of the $x = 0$ axis. The figure is generated by a pseudo-random-number generator, and the step length is ± 1, giving rise to the horizontal 'stripes' in the figure, with separations of 1 unit.

Obtain an expression for the mean-square displacements of the particle $\langle X^2 \rangle$ and $\langle Y^2 \rangle$ in the limit of a large number of steps n.

In spite of its simplicity, the random-walk model contains many important features of Brownian motion. In particular, it serves to explain one of the arguments against the hypothesis that molecular bombardment is the origin of Brownian motion. Very early on, von Nägeli (1879) had considered this as a possible explanation, but rejected the idea after having esti-mated the momentum a *single* atomic collision might impart to a macroscopic Brownian particle (MacDonald, 1962). The change in velocity of a particle due to one such collision would be $\delta V \simeq u(m/M)$, where u is the velocity of a molecule with mass m, and M is the mass of the Brownian particle. Assume as an order of magnitude that $m \approx 10^{-22}$ g and a typical thermal velocity $u \approx 2 \times 10^4 \, \text{m s}^{-1}$, corresponding to a temperature of 300 K. With $M \simeq 10^{-12}$ g we find $\delta V \simeq 10^{-6} \, \text{cm s}^{-1}$. The observed displacements for actual instances of Brownian motion were found to correspond to velocities that were estimated to be 100 times larger, and von Nägeli eventually abandoned the whole idea. His arguments were, however, implicitly assuming that the impacts from the molecules are 'alternating' more or less regularly, so to speak. We have seen that this is not so; there are long intervals of time during which the particle is moving in essentially the same direction, meaning that many of the small impacts add up for extended periods of time. All the foregoing arguments are circumvented by appreciating from the very beginning that even a macroscopic particle can be considered as a giant molecule and its mean-square velocity calculated from first principles when it is in thermal equilibrium with a surrounding fluid (MacDonald, 1962).

- **Example:** An elastically bound particle undergoing a random walk can be described by assigning the step probabilities $\frac{1}{2}(1 + k/R)$ and $\frac{1}{2}(1 - k/R)$ for moving to the left or to the right, respectively, provided that the particle is at the position labeled k, see, e.g., Kac (1946). Here, R is a certain integer, and the possible positions of the particle are limited by the condition $-R \le k \le R$. Every time the particle reaches the position $+R$ or $-R$ it returns with probability 1.

A result for this case is shown in Fig. 7.4, to be compared with Fig. 7.3, taking $R = 5$. Note that now all positions are restricted to the interval $-5 \le x \le 5$. Although the particle is elastically bound, it does not actually oscillate in x; every time it reaches the origin, $x = 0$, it has the same probability of stepping in each direction; hence there will not be any regular frequency for the zero-crossings. Figure 7.4 has the appearance of the motion of an overdamped harmonic oscillator, driven by random noise. The oscillatory motion originates from inertial effects that are absent from the simple description of Brownian motion, as mentioned in Section 6.1.1. The inertial effects cause the particle to continue, at least for a little while, past the position $x = 0$, until the elastic force turns it back.

- **Example:** The diffusion problems discussed in the present treatise are basically of the classical type, for which mean-square displacements of particles are, at least asymptotically, proportional to time. More generally we have $\langle r^2 \rangle = Bt^\alpha$ as $t \to \infty$, where α is the diffusion exponent (Balescu, 1997).

The case with $\alpha = 1$ is the normal diffusive range. The 'ballistic' case corresponds to $\alpha = 2$, and can be realized by particles in free flight, i.e. in each realization we have $X = Ut$, with U constant in each individual realization. The ensemble average becomes $\langle X^2 \rangle = \langle U^2 \rangle t^2$. The case with $\alpha < 1$ is the subdiffusive case. The case $1 < \alpha < 2$ is the superdiffusive case. Often we see the term 'anomalous transport', in particular in relation to turbulent transport, although this terminology is not unambiguous. It might be argued that 'strange' diffusion is a more appropriate term (Balescu, 1997).

An interesting example realizing 'strange' diffusion has been described by Dreizin and Dykhne (1972, 1973). They consider a particle in random walk in the y-direction on a plane divided into 'stripes' of equal width, all being parallel to the x-axis. The width, ℓ, of the stripes is much wider than the step-length of the random walk. Assume that the 'stripes' are moving with constant velocity U in the x-direction,

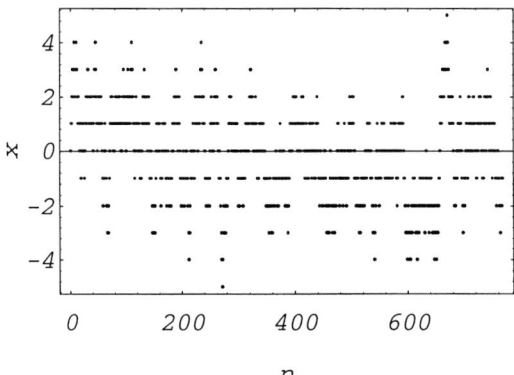

Figure 7.4 A single realization of an extended random walk with 750 steps for a harmonically bound particle. The figure is generated by a pseudo-random-number generator, as in Fig. 7.3, and the step length is ± 1 here also, giving rise to the horizontal 'stripes' in the figure, with separations of 1 unit. In this calculation we assumed that the parameter $R = 5$.

and that the *direction* of propagation is chosen randomly for each individual stripe, independent of the motion of the others. If a particle remains at rest in a time interval of duration t, it is thus displaced by a distance Ut in the x-direction.

The x-displacement of the particle as it migrates by its random motion along the y-axis is thus $X \approx Ut\,\Delta N/N$, where $\Delta N/N$ is the fraction of 'uncompensated' layers. If the particle has passed equally many stripes with velocities in the positive and in the negative x-direction we have $\Delta N = 0$. Since the motion in the y-direction was assumed to be diffusive with a diffusion coefficient D, we have $N \approx \sqrt{Dt}/\ell$. Since we have assumed that the direction of motion of the stripes can take either value with equal probability, and that this direction is chosen randomly for the individual stripes, we have $\Delta N \approx \sqrt{N}$. The result for the mean-square displacement of the particle in the x-direction can therefore be estimated heuristically as $\langle X^2 \rangle \approx t^{3/2} U^2 \ell / \sqrt{D}$. This result is only approximate, because we have *estimated* rather than actually *calculated* the average $\langle (\Delta N/N)^2 \rangle$.

7.1.1   Reflecting barriers

An important generalization of the foregoing results can be obtained by including the effects of barriers. Here we consider two types, a perfect reflector and a perfect absorber.

Consider first an ideally reflecting barrier; by this we mean one having the property that, if a particle undergoing a one-dimensional random walk arrives at the barrier at position m_r, it will with certainty be reflected one step back at its next timestep, i.e. it will then arrive at the position with label $m_r - 1$. We now obtain an expression for the probability $P_{r,n;m_r}$ of finding a particle at the position labeled r at time labeled n, given that a reflecting barrier is placed at position m_r, as indicated by the subscript r. Two basically different situations may arise; either the particle has arrived at r without undergoing any reflections, or it has bounced off the barrier one or more times before arriving at r. In the latter case we can construct a virtual path by a mirror reflection of the actual path after reflection with respect to the position m_r, as shown in Fig. 7.5.

We will now express $P_{r,n;m_r}$ in terms of the probability $P_{r,n}$, see (7.2), which refers to the case in which the barrier is absent. Consider first a path undergoing only one reflection at m_r.

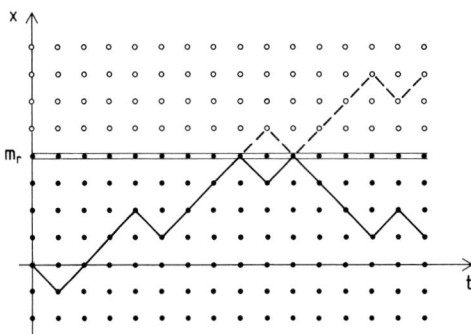

Figure 7.5 An illustration of the space–time variation of a simple one-dimensional random walk in the presence of a reflecting barrier at the position m_r. The dashed line indicates a virtual path that is explained in the text. Positions on the time axis have the label n.

By constructing the corresponding virtual path, we find that this will lead to the image position $2m_r - r$ at time n. Conversely, in the absence of a barrier, for every trajectory leading to the image point having crossed the line m_r only once, there is exactly one that leads to m_r after a single reflection. We can argue in a similar way for a trajectory undergoing *two* reflections. In this case the actual path can, in the absence of a barrier, be completed with a segment of the mirror path between the two reflections, and this path will still end at (r, n). In the case of three reflections the actual path can, in the absence of a barrier, be replaced by one having a mirror segment between the two reflections, and then again a mirror segment from the third reflection until it reaches the point $(2m_r - r, n)$. In this way we can uniquely present each path undergoing reflections in the case *with* a barrier, with another path *in the absence* of a barrier. All the relevant paths with zero or an even number of reflections end at (r, n), and these are represented by a probability $P_{r,n}$. All those with uneven numbers of reflections end at $(2m_r - r, n)$ and these are represented by $P_{2m_r-r,n}$. Consequently we have (Chandrasekhar, 1943)

$$P_{r,n;m_r} = P_{r,n} + P_{2m_r-r,n}, \qquad r < m_r, \tag{7.5}$$

relating the probability *with* a reflecting barrier present at m to the corresponding probabilities *without* such a barrier. Note in particular that $P_{m_r,n;m_r} = 2P_{m_r,n}$, i.e. *with* a reflecting barrier we have twice the probability of finding a particle at the barrier position that we have in the case in which no such barrier is present. This result can be understood on intuitive grounds.

7.1.2   Absorbing barriers

As an alternative to the discussion in the foregoing subsection, we here consider a perfectly absorbing barrier at the position m_a. By a perfect absorber we here mean one with the property that, if a particle undergoing a one-dimensional random walk arrives at the barrier, it will disappear from the system with certainty. The problem of 'gambler's ruin' mentioned before can be formulated in this way; the absorbing barrier can represent the funds available to the gambler, and, in the likely case that no credit is offered, it represents the maximum loss he can suffer (Cox and Miller, 1965).

Also in the case of absorbing barriers we have two possibilities; a particle arrives at r at time n without ever reaching a barrier, or it can arrive at m_a at an earlier time and be lost from the system before reaching r. Once again we introduce the virtual path discussed in Section 7.1.1 and Fig. 7.5. The present problem can be defined as one in which we take *all* paths that arrive at r at time n without the presence of a barrier, and remove from these all those which at some time encountered the barrier position m_a at least once. For these we note that, to every path arriving *at* the barrier, there correspond uniquely defined paths that, in the absence of the barrier, lead to the point $2m_a - r$ at time n. The probability of finding a particle at the position r at time n with an absorbing barrier present at position m_a is consequently

$$P_{r,n;m_a} = P_{r,n} - P_{2m_a-r,n}, \qquad r < m_a. \tag{7.6}$$

The expression (7.6) relates the probability *with* an absorbing barrier present at m_a to the corresponding probabilities *without* such a barrier. Note in particular that $P_{m_a,n;m_a} = 0$, as expected.

Recall that $P_{r,n} = 0$ for $r > n$, implying that also $P_{2m_a-r,n} = 0$ for $r > n$ with $n < m_a$. Within this time the particle has no possibility of reaching the position r and the probability $P_{r,n;m_a} = 0$, which is as it should be.

7.1.3   The probable rate of arrival

The result for $P_{r,n;m_a}$ obtained in Section 7.1.1 expresses the probability of actually finding a particle at the position r at time n with an absorbing barrier present at position m_a. In the derivation all orbits touching the barrier itself at some time were excluded, and the result $P_{m_a,n;m_a} = 0$ is therefore somewhat trivial. A different and equally meaningful question is to ask what the rate of arrival $a_{m_a,n}$ of particles *at* the absorbing barrier is. For this purpose we now assume that a particle *has* arrived at the screen position m_a at time n, without ever having touched that position before. Note that the meaning of $a_{m_a,n}$ is different than $P_{r=m_r,n;m_r}$ and $P_{r=m_a,n;m_a}$.

It is evident for the present analysis that n has to be even (odd) if m_a is even (odd), respectively. We assume that this is the case. We now attempt again to express $a_{m_a,n}$ in terms of $P_{r,n}$ relating to the conditions without the presence of the barrier. On removing the barrier, we note that a particle can arrive at (m_a, n) either from $(m_a - 1, n - 1)$ or from $(m_a + 1, n - 1)$. The latter position is forbidden in the presence of the barrier, and we want to remove it from the allowed paths. This is, however, not sufficient, since also some of the paths leading to $(m_a - 1, n - 1)$ are forbidden, since they touch the position m_a once or more at earlier times. To identify these, we note, by considering the mirror images of the orbits with respect to the line $r = m_a$ in an r–n plane, that to every orbit passing through $(m_a + 1, n - 1)$ we can uniquely assign another orbit of forbidden nature arriving at $(m_a - 1, n - 1)$ (forbidden in the sense that it touches or crosses the line $r = m_a$ at some instant). Thus, all in all, the number of permitted ways of arriving at the position m_a for the first time after n steps is equal to all the possible ways of arriving at m_a after n steps in the absence of the absorber minus *twice* the number of ways of arriving at $m_a - 1$ after $n - 1$ steps again in the absence of the absorber. We therefore have

$$\frac{n!}{[\frac{1}{2}(n - m_a)]![\frac{1}{2}(n + m_a)]!} - 2\frac{n!}{[\frac{1}{2}(n + m_a)]![\frac{1}{2}(n - m_a - 2)]!}$$

$$= \frac{n!}{[\frac{1}{2}(n - m_a)]![\frac{1}{2}(n + m_a)]!}\left(1 - \frac{n - m_a}{n}\right)$$

$$= \frac{n!}{[\frac{1}{2}(n - m_a)]![\frac{1}{2}(n + m_a)]!}\frac{m_a}{n} \qquad (7.7)$$

permitted paths. Recalling again that there are 2^n paths with duration n, we obtain that the probability of a particle arriving for the first time at m_a after exactly n steps is

$$a_{m_a,n} = \frac{m_a}{n}P_{m_a,n}, \qquad r < m_a, \qquad (7.8)$$

with $P_{m_a,n}$ defined by (7.2). Evidently, $a_{m_a,n} < P_{m_a,n}$ for all $n > 1$.

- **Exercise:** The symmetry arguments in the last three subsections may appear deceivingly simple and convincing; the reader is strongly urged to convince himself/herself that they are indeed complete.

7.1.4   Limits of continuous distributions

By letting N be large we can apply a coarse-grained description whereby the relevant probabilities are considered to be continuous functions. The results of Sections 7.1.1, 7.1.2, and 7.1.3 will be considered independently.

7.1.4.1   The case of no barriers

For later reference, it is advantageous first to consider the case in which there are no barriers present, to obtain the results for the continuum limit. We first take the appropriate result (7.2) in the limit of large n and use Stirling's formula in the form

$$\log n! = \left(n + \tfrac{1}{2}\right)\log n - n + \tfrac{1}{2}\log(2\pi) + \mathcal{O}(n^{-1})$$

to approximate

$$P_{r,n} = \frac{n!}{[\tfrac{1}{2}(n-r)]![\tfrac{1}{2}(n+r)]!}\left(\frac{1}{2}\right)^n . \tag{7.9}$$

Obtaining

$$\log P_{r,n} \approx -\frac{1}{2}\log n + \log 2 - \frac{1}{2}\log(2\pi) - \frac{r^2}{2n}, \tag{7.10}$$

with use of $\log(1 \pm r/n) \approx \pm(r/n) - [r^2/(2n^2)]$, we readily find (Chandrasekhar, 1943) that

$$P_{r,n} \approx \left(\frac{2}{\pi n}\right)^{1/2}\exp\left(-\frac{r^2}{2n}\right). \tag{7.11}$$

Inroducing the net displacement as $x = r\Delta_x$ and the actual time as $t = n\Delta_t$, and defining $D = \tfrac{1}{2}n\Delta_x^2$, we rewrite $P_{m,n}$ as a function of new variables:

$$P(x, t) \approx \frac{1}{2(\pi t D)^{1/2}}\exp\left(-\frac{x^2}{4Dt}\right). \tag{7.12}$$

The probability of a particle being found in the interval $\{x, x + \Delta\}$ at a time t is expressed as $P(x, t)\Delta$. Note that (7.12) is a solution to the diffusion equation

$$\frac{\partial}{\partial t}P(x, t) = D\,\frac{\partial^2}{\partial x^2}P(x, t),$$

where D is a diffusion coefficient, and the initial condition is a δ-function (see also Appendix D), $P(x, t = 0) = \delta(x)$, i.e. the particle is known with certainty to be at the origin at time $t = 0$.

7.1.4.2   The case of reflecting barriers

Using the result (7.11), we find that (7.5) can be approximated as

$$P_{r,n;m_r} \approx \left(\frac{2}{\pi n}\right)^{1/2}\left[\exp\left(-\frac{m^2}{2n}\right) + \exp\left(-\frac{(2m_r - m)^2}{2n}\right)\right] \tag{7.13}$$

for $n \to \infty$. By introducing the net displacement as $x = r\Delta_x$ and the actual time as $t = n\Delta_t$, and defining $D = \tfrac{1}{2}n\Delta_x^2$, we rewrite $P_{r,n;m_r}$ as a function of new variables

$$P_r(x, t) \approx \frac{1}{2(\pi t D)^{1/2}} \left[\exp\left(-\frac{x^2}{4Dt}\right) - \exp\left(-\frac{(2x_r - x)^2}{4Dt}\right) \right], \tag{7.14}$$

x_r being the position of the reflecting barrier. From (7.14) we find $\partial P_r(x, t)/\partial x|_{x=x_r} = 0$, without conditions on the actual value of $P_r(x, t)$ at the boundary.

7.1.4.3   The case of absorbing barriers

Using the results (7.9)–(7.11), we find that (7.6) can be approximated by

$$P_{r,n;m_a} \approx \left(\frac{2}{\pi n}\right)^{1/2} \left[\exp\left(-\frac{r^2}{2n}\right) - \exp\left(-\frac{(2m_a - r)^2}{2n}\right) \right] \tag{7.15}$$

as $n \to \infty$. Introducing again the net displacement as $x = r\Delta_x$, the actual time as $t = n\Delta_t$, and D as defined before, we rewrite $P_{r,n;m_a}$ as a function of new variables:

$$P_a(x, t) \approx \frac{1}{2(\pi t D)^{1/2}} \left[\exp\left(-\frac{x^2}{4Dt}\right) - \exp\left(-\frac{(2x_a - x)^2}{4Dt}\right) \right], \tag{7.16}$$

x_a being the position of the absorbing barrier. From (7.16) we find that $P_a(x = x_r, t) = 0$, without conditions on the value of the spatial derivative, $\partial P_r(x, t)/\partial x = 0$, at the boundary.

7.1.4.4   The probable rate of arrival

For the limiting case of large n we obtain from (7.8) and by use of the foregoing results

$$a_{m_a,n} \approx \frac{m_a}{n} \left(\frac{2}{\pi n}\right)^{1/2} \exp\left(-\frac{m_a^2}{2n}\right), \tag{7.17}$$

or, in terms of the variables x_a and t introduced before,

$$a(x_a, t) \approx \frac{x_a}{nt} \frac{1}{(\pi t D)^{1/2}} \exp\left(-\frac{x_a^2}{4Dt}\right). \tag{7.18}$$

Finally we ask for the probability $q(x_a, t)\Delta$, that the particle arrives at x_a in the time interval t, $t + \Delta$ for the first time. We have $q(x_a, t)\Delta = \frac{1}{2} a(x_a, t)n\Delta$, since $a(x_a, t)$ is the number of particles arriving at x_a in the time it takes to traverse *two* steps, see Section 7.1.3. We find that

$$q(x_a, t) \approx \frac{x_a}{t} \frac{1}{2(\pi t D)^{1/2}} \exp\left(-\frac{x_a^2}{4Dt}\right). \tag{7.19}$$

We can interpret (7.19) as giving the fraction of a large number of particles released initially at $x = 0$ that is deposited on an absorbing screen per unit time at the time t, assuming that the particles execute their random walks independently of each other.

It is easily verified that $q(x_a, t)$ satisfies the relation

$$q(x_a, t) = -D \frac{\partial}{\partial x} P_a(x, t)\Big|_{x=x_a},$$

with $P_a(x, t)$ given by (7.16). This result can be understood by recalling that $[1/(2\sqrt{\pi t D})]$ $\exp[-x^2/(4Dt)]$ is a solution to the diffusion equation

$$\frac{\partial}{\partial t} P(x, t) - \frac{\partial}{\partial x} D \frac{\partial}{\partial x} P(x, t) = 0, \tag{7.20}$$

where we placed the constant diffusion coefficient, D, under the first differentiation sign for later reference. On comparing this form of the diffusion equation with the equation of continuity written in one spatial dimension,

$$\frac{\partial}{\partial t} \rho(x, t) + \frac{\partial}{\partial x} u(x, t)\rho(x, t) = 0,$$

for a fluid with local density ρ and velocity u, we see that $-D\, \partial P(x, t)/\partial x$ in the diffusion equation has the same role as that of the particle-flux density $u(x, t)\rho(x, t)$ in the continuity equation, implying that we may interpret $-D\, \partial P(x, t)/\partial x$ as a 'probability flux.'

7.1.5   Random walks in two and three spatial dimensions

The foregoing discussions refer to one spatial dimension. The dimensionality of the random-walk process being considered is not trivial. In one spatial dimension it is easily realized that, in the limit $t \to \infty$, every spatial position on the discretized x-axis will be visited at least once, with probability 1; the process is said to be *recurrent*. In two spatial dimensions the answer is no longer evident, but it can be demonstrated (Chung and Fuchs, 1951, Cox and Miller, 1977) that every spatial position in the discretized x–y plane will, with probability 1, be visited at least once. In three spatial dimensions this is no longer so! In a three-dimensional random walk, no spatial position is recurrent. This distinction in the spatial dimensionality is lost in the continuum limit.

7.2   Lévi flights

The foregoing discussions of random walks assumed that the position is, or at least can in principle be, identified at each timestep. In reality this need not be the case; for instance it is quite impossible to determine the position of a grain of pollen at each impact from the surrounding medium during its Brownian motion. Rather it is a position as $X = X_1 + X_2 + \cdots + X_N$ that is observed, where X_i with $i = 1, 2, \ldots, N$ are the individual small displacements. In the random walk discussed previously $X_i = \pm\Delta_x$. In general, we allow X_i to be statistically distributed, with the probability density $P(X_i)$ assumed known for all i. In principle also N can be a random variable, i.e. there need not be the same number of individual small displacements between successive observations of positions. For simplicity, we shall ignore this possibility in the following.

The obvious question that now arises is this: in what sense is our interpretation of the observations influenced by the 'stroboscopic' sampling of the process when only $P_N(X_i)$ is known, not the actual value of N? In other words, under which conditions are there simple relations between the statistics of X and the individual X_i? In more philosophical terms (Klafter et al., 1996), when does the whole look like the part? When X_i is statistically independent of X_j for $j \neq i$, this question is evidently analyzed best in terms of the characteristic functions. A simple and important case is the one in which $P(X_i)$ is a Gaussian for all i. Then the respective characteristic functions are Gaussians, and the probability density $P_N(X)$ is also a Gaussian with a variance equal to the sum of the individual variances. Although this example can be assumed to be the most relevant one for the classical problem of Brownian motion, it is not the only one. Cauchy had noticed already in 1853 that, if the probability density $P(X) = (1/\pi)(1 + x^2)^{-1}$, then $P_N(X) = [1/(\pi N)][1 + (x/N)^2]^{-1} = (1/N)P(X/N)$, known as the Cauchy distribution (Klafter *et al.*, 1996, Balescu, 1997, Reichl, 1998). More general results were obtained by the French mathematician Paul Lévi (1937) who considered characteristic functions of the form

$$CH_N(k) = e^{N|k|^{\beta}}, \tag{7.21}$$

with β constant. The characteristic function (7.21) is the product of N individual functions $\exp(-|k|^{\beta})$, with examples shown in Fig. 7.6. The inverse Fourier transform of $CH_N(k)$ then gives the probability density P_N. Since $CH_N(k = 0) = 1$, the probability density $P_N(X)$ is automatically normalized for all $\beta > 0$, but Lévi demonstrated that it is necessary that $0 < \beta \le 2$ in order to ensure that $P_N(X) \ge 0$. The proof will not be given here, but it seems reasonable enough; it can be seen from Fig. 7.6 that, as β increases, $\exp(-|k|^{\beta})$ approaches the box function which is 1 for $|k| < 1$ and 0 for $|k| > 1$. The probability density obtained by Fourier transformation of this function is, however, known to be $\sin(x)/x$, which assumes negative values. Consequently the box function and its like are not acceptable as characteristic functions; they are not non-negative definite, see also Fig. 7.7.

By an asymptotic series expansion (Lighthill, 1964), see also Section 2.6, it can be demonstrated that, for large $|x|$, the probability density $P_N(X)$ varies approximately as $1/|x|^{1+\beta}$. Consequently the second moment of $P_N(X)$ diverges for $\beta < 2$ and the distribution possesses no variance and therefore no characteristic step length $\langle x^2 \rangle^{1/2}$ can be defined. In the Gaussian case with $\beta = 2$ in (7.21) we have a finite value for $\langle x^2 \rangle$, in which case the Brownian random walk results. On the other hand, for $\beta \to 0$ the characteristic function (7.21) approaches the constant $1/e$ with a spike around the origin added. The constant contributes with a delta-function to the probability density and the spike with a long tail, which gives rise to the divergence of $\langle x^2 \rangle$.

- **Example:** As an illustration of some of the properties of Lévi processes consider a simple one-dimensional case with step sizes $1, b, b^2, b^3, \ldots$, with a probability density for the step length x of

$$P(x) = \frac{\lambda - 1}{2\lambda} \sum_{n=0}^{\infty} \frac{1}{\lambda^n} (\delta_{x,b^n} + \delta_{x,-b^n}), \tag{7.22}$$

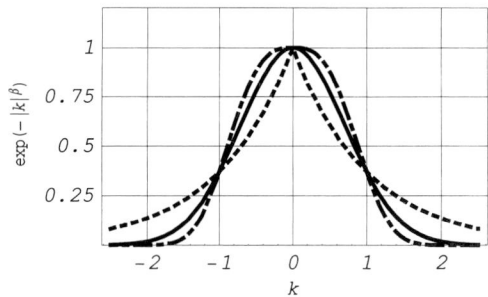

Figure 7.6 Examples of characteristic functions $\exp(-|k|^{\beta})$ for various values of β. The dashed line corresponds to $\beta = 1$, the full line to $\beta = 2$, and the dot–dashed line to $\beta = 3$.

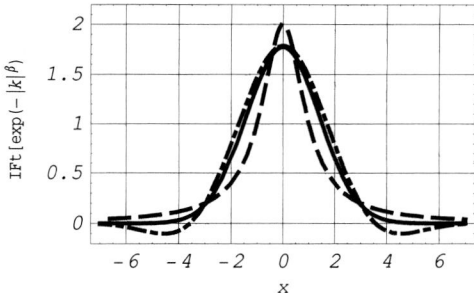

Figure 7.7 The probability densities obtained by inverse Fourier transformation of the characteristic functions IFt[exp($-|k|^\beta$)] shown in Fig. 7.6 for various values of β. The dashed line corresponds to $\beta = 1$, the full line to $\beta = 2$, and the dot–dashed line to $\beta = 3$. Note that the probability densities for $\beta > 2$ assume negative values and are therefore unacceptable. The function for $\beta = 3$ has a rather complicated analytic expression in terms of hypergeometric functions. There is little point in presenting this explicitly here.

where $\lambda > 1$ and $b > 1$ are parameters characterizing the process. The Kronecker δ was introduced. There is no particular mathematical reason for b being larger than 1, but the physical appearance of the process changes drastically for $b < 1$. Evidently the probability density (7.22) has the property that, every time the step length is increased by a factor b, the probability of that step length decreases by a factor of $1/\lambda$. Starting the process at the origin, one finds typically a cluster of λ steps each having the length 1 before there is a jump of length b^2 and the process usually remains near the new position for some time. After a while also a jump to a position at a distance b^3 will occur, and the following points are likely to cluster in its vicinity. The process continues, forming a hierarchy of clusters within clusters (Klafter *et al.*, 1996, Reichl, 1998). The Fourier transform of (7.22) is

$$CH(k) = \frac{\lambda - 1}{2\lambda} \sum_{n=0}^{\infty} \frac{1}{\lambda^n} \cos(b^n k), \tag{7.23}$$

which is known as the Weierstrass function.

Note the self-similarity of (7.23). To make this explicit, (7.23) is rewritten as

$$CH(k) = \frac{1}{\lambda} CH(bk) + \frac{\lambda - 1}{\lambda} \cos(bk). \tag{7.24}$$

For small k, this equation has the approximate solution

$$CH(k) = e^{-|k|^\beta}, \tag{7.25}$$

with $\beta = \ln \lambda / \ln b$ making the relation to the Lévi problem self-evident. The equivalence to (7.21) is not complete since (7.25) is only an approximation; evidently, if we zoom in on the random walk described by (7.22), its smallest scale or smallest step length is noticeable, but, in a coarse-grained visualization, its properties will appear to be reproduced on large as well as small scales.

Lévi random walks or Lévi flights have been used to model certain types of random-walk processes observed in quasi-turbulent flows (Solomon *et al.*, 1993).

8 Density fluctuations in gases

A consequence of the random motion of the molecules of a gas is that their number per unit volume fluctuates about the average value. As the simplest illustration of this point we consider a large volume V_0 containing N_0 molecules. Selecting (at random) a small partial volume $V \ll V_0$, it is now assumed that the probability of finding a selected molecule in V is given by

$$P = \frac{V}{V_0},$$

independent of the distribution of all other molecules, and independent of the actual shape of the volume V. It is implicitly assumed that, by virtue of their thermal motion, any particle eventually scans the entire available volume. On the basis of this *a priori* probability it is readily demonstrated that the number N of molecules in V has a binomial distribution, see Appendix A. The probability of finding exactly N molecules in the volume V is thus

$$P(N \text{ molecules in } V) = \frac{N_0!}{N!(N_0 - N)!}\left(\frac{V}{V_0}\right)^N\left(1 - \frac{V}{V_0}\right)^{N_0 - N}.$$

In the limit $N_0 \to \infty$ and $V_0 \to \infty$ so that the density $N_0/V_0 \equiv \mu_0$ is a constant, we find the average value $\langle N \rangle = \mu_0 V$ and the standard deviation $\langle N^2 \rangle = \mu_0 V = \langle N \rangle$. The *relative* fluctuation is

$$\frac{\sqrt{\langle N^2 \rangle}}{\langle N \rangle} = \frac{1}{\sqrt{\langle N \rangle}}. \tag{8.1}$$

This result can be expressed in terms of standard quantities, namely pressure, p, and temperature. With $p = n\kappa T$, we have $p = \kappa T \langle N \rangle / V$, giving

$$\frac{\sqrt{\langle N^2 \rangle}}{\langle N \rangle} = \sqrt{\frac{\kappa T}{pV}}.$$

Although it is seemingly simple, this relation is of little use as it stands; V is the subvolume under consideration, $V \ll V_0$, and the meaning of this symbol is well defined. On the other hand, p and T are in fact not yet determined. It is evident, however, that, in the present context, the subvolume is in complete equilibrium with the full system, there are no *physical* walls separating the subvolume V from the rest, and particles are assumed to flow freely through the subvolume. By definition we therefore have $p = p_0$ and $T = T_0$ with the subscript indicating the complete system with volume V_0, i.e. $p_0 = \kappa T_0 N_0 / V_0$. Consequently

$$\frac{\sqrt{\langle N^2 \rangle}}{\langle N \rangle} = \sqrt{\frac{\kappa T_0}{p_0 V}}, \tag{8.2}$$

which states that the relative fluctuation in number of particles vanishes as V increases, as expected. A small volume having linear dimensions of the order of the wavelength of blue light $(450\text{–}500 \text{ nm} = (4.5\text{–}5.0) \times 10^{-7} \text{ m})$ contains about 3×10^6 molecules, and, in such a volume, $\sqrt{\langle N^2 \rangle}/\langle N \rangle \approx 5.7 \times 10^{-4}$, or 0.057%. The corresponding values for red light (610–780 nm) are $\sqrt{\langle N^2 \rangle}/\langle N \rangle \approx 3.5 \times 10^{-4}$, or 0.035%. Fluctuations of this order of magnitude result in variations of the index of refraction throughout a gas of sufficient magnitude to cause appreciable scattering of light. As the numerical values presented previously indicate, the effect is, however, more pronounced for blue light than it is for red; blue light is scattered more than is red, which, for instance, accounts for the blue color of the sky.

For small volumes, V, of the order of ℓ^3, where ℓ is the average interparticle separation, expressions (8.1) and (8.2) are expected to be quite accurate. However, for larger volumes the basis of the derivation is questionable for actual physical systems. Thus, if by chance the number of molecules in V is large, an additional incoming molecule is likely to be pushed away if we assume the interaction between molecules to be as for billiard-ball collisions. In this limit it can no longer be assumed that the presence of one molecule in V is independent of the position of all the others. The details of the corresponding analysis will evidently depend on the nature of the actual interaction. In that which follows, this problem is addressed on the basis of the exposition of Tolman (1938) and Lee et al. (1973), although the formulation they used had been proposed already by Einstein (1956). Some details in the derivation are highly idealized but parts are, on the other hand, kept in rather general terms so that the results can be illustrated by specific examples.

We consider, in our imagination, a very large volume, V_0, of the gas subdivided into a large number N of smaller volume elements, V, of subsystems. The remainder of the gas serves as the heat reservoir for any one subsystem. Each subsystem contains a large number of molecules, n. (We here use n for the number of molecules per subsystem rather than for the number per unit volume.) For example, we might consider a volume of 10 l (10,000 cm^3) of a gas and imagine it subdivided into subsystems of volume elements, 1 mm^3 each. Then there are $N = 10^7$ subsystems, and, under standard conditions, $n = 3 \times 10^{16}$ molecules per subsystem.

Each subsystem is considered so large that the penetration depth of a molecule coming from the outside can be taken to be negligibly small compared with the size of the subsystem. The impact of such an 'external' molecule on the surface is, however, inelastic in the sense that energy transferred to molecules in the outer layer is distributed among the molecules in the subsystem. Thus, in effect, we consider each subsystem of molecules to be enclosed within a perfectly flexible heat-conduction sack, so that there is negligible exchange of molecules between systems. To substantiate this model, we estimate the typical collisional mean free path by taking the characteristic atomic diameter to be $a \simeq 1.5 \times 10^{-10}$ m, giving a cross-section of $\sigma_c \simeq 3 \times 10^{-20}$ m^2. The collisional mean free path is then $\ell_c \simeq 1/(\sigma_c \eta)$, η here being the number density of atoms. With $\eta \approx 3 \times 10^{25}$ m^{-3} we find $\ell_c \simeq 10^{-6}$ m. This distance is negligible compared with the subsystem sizes of 1 mm under consideration here. Consequently, on average, it takes many collisions for a molecule to migrate from one subsystem to another.

As a result of random molecular motions, the molecules of any one subsystem will at times occupy a volume somewhat smaller than the normal volume V_0, and at other times a volume somewhat greater. At any one instant there will be fluctuations in density on going from subsystem to subsystem throughout the body of the gas or, if we follow any one small sub-system as time passes, its density will fluctuate with time.

As the basic assumption of the following analysis we have that the number of subsystems with a volume in the interval $\{V; V + dV\}$ is

$$dN_V = \alpha \exp[-\phi/(\kappa T)]\, dV, \tag{8.3}$$

where α is a constant to be determined and $\phi = \phi(V)$ is a potential energy, which will be expressed in terms of thermodynamic quantities. The relation (8.3) will here be postulated directly, but it can easily be derived from the full energy distribution for the particles in the gas, see for instance Chapter 11.

Without loss of generality we are free to consider the potential energy of a system zero when it has the volume V_0 and pressure p_0 appropriate to a perfect uniform distribution of molecules throughout the main body of the gas. The potential energy in any other state, of volume V and pressure p, is the work done on the system to bring it from the uniform state to this new state. We assume that, during the change, the remainder of the gas exerts the pressure p_0 on the system. Then, if p is the pressure exerted by the molecules of the system at any instant, the net external pressure is $p_0 - p$, and the work done on the system while its volume changes from V_0 to V, or the potential energy ϕ, is

$$\phi = \int_{V_0}^{V} (p_0 - p)\, dV = -\int_{V_0}^{V} (p - p_0)\, dV. \tag{8.4}$$

The pressure p is some function of the volume V. Without making any special assumption regarding the equation of state of the gas, we use a Taylor series to express the pressure p as

$$p = p_0 + \left(\frac{\partial p}{\partial V}\right)_0 (V - V_0) + \left(\frac{\partial^2 p}{\partial V^2}\right)_0 \frac{(V - V_0)^2}{2!}$$
$$+ \left(\frac{\partial^3 p}{\partial V^3}\right)_0 \frac{(V - V_0)^3}{3!} + \cdots, \tag{8.5}$$

where the derivatives are to be evaluated at constant temperature and for the volume V_0.

8.1   Ideal gases

We consider first the special case of an ideal gas, for which $pV = n\kappa T$, where n is the number of molecules in V. For this case we have

$$\left(\frac{\partial p}{\partial V}\right)_0 = -\frac{1}{V_0^2} n\kappa T,$$

$$\left(\frac{\partial^2 p}{\partial V^2}\right)_0 = \frac{2}{V_0^3} n\kappa T,$$

etc., and, since from the nature of the problem the fluctuations are small and $\Delta V \equiv V - V_0 \ll V_0$, it follows from (8.5) that the terms containing the second and higher derivatives are very small compared with that containing the first derivative. Then we can write

$$p - p_0 = -n\kappa T \frac{V - V_0}{V_0^2}$$

$$= -p_0 \frac{\Delta V}{V_0}. \tag{8.6}$$

The fractional change in volume, $\Delta V/V_0$, was called the *condensation* δ by Smoluchowski (*Verdichtung* in the German original). The notation is standard, and should be confused neither with the Kronecker δ nor with Dirac's δ-function. It can be written as

$$\delta = \frac{\Delta V}{V_0} = \frac{V - V_0}{V_0} = \frac{V}{V_0} - 1. \tag{8.7}$$

Consequently, $d\delta = dV/V_0$, and using (8.6), we can write (8.4) in the form

$$\phi = \int_0^\delta p_0 V_0 \delta \, d\delta$$

$$= \frac{p_0 V_0 \delta^2}{2} = \frac{n\kappa T \delta^2}{2}, \tag{8.8}$$

giving $\phi/(\kappa T) = n\delta^2/2$. For an ideal gas, equation (8.3) therefore becomes

$$dN_\delta = V_0 \alpha \exp\left(-\tfrac{1}{2} n \delta^2\right) d\delta. \tag{8.9}$$

The constant α is determined from the requirement that the integral of dN_δ over all possible values of δ must equal the total number of systems, N. Now δ is always a small quantity, but let us suppose that it is as large as $\pm 10^{-2}$. Then $\delta^2 = 10^{-4}$. Also, suppose that we consider a system so small that it contains only 2×10^6 molecules (in general, n is much larger than this). Then $n\delta^2/2 = 100$ and $\exp(-n\delta^2/2) = \exp(-100) \approx 4 \times 10^{-44}$. Hence the exponential term is already so small when $\delta = \pm 10^{-2}$ that we might as well, without appreciable error, take the limits of integration from $-\infty$ to ∞. Thus

$$N = \int dN_\delta = V_0 \alpha \int_{-\infty}^\infty \exp\left(-\frac{n\delta^2}{2}\right) d\delta$$

$$= V_0 \alpha \sqrt{2\pi/n},$$

and consequently $V_0 \alpha = N\sqrt{n/(2\pi)}$. Then, from (8.9),

$$\frac{dN}{N} = \sqrt{\frac{n}{2\pi}} \exp\left(-\frac{n\delta^2}{2}\right) d\delta. \tag{8.10}$$

The ratio dN/N can be considered either as the fractional number of systems for which the condensation, at any instant, lies between δ and $\delta + d\delta$, or as the fraction of time spent by any one system in this state. Note that the number of systems in which the condensation has the value δ at any instant decreases exponentially both with n and with δ^2. The probability density $P(\delta)$ is given by $\exp(-n\delta^2/2)$, apart from a constant that can be determined by requiring that $\int_{-\infty}^\infty P(\delta) \, d\delta = 1$.

The average value of the condensation $\langle \delta \rangle$ is evidently zero. The root-mean-square condensation, $\langle \delta^2 \rangle^{1/2} \equiv \delta_{rms}$, is

$$\delta_{rms} = \left(\frac{\int_{-\infty}^\infty \delta^2 \exp(-n\delta^2/2) \, d\delta}{\int_{-\infty}^\infty \exp(-n\delta^2/2) \, d\delta}\right)^{1/2} = \frac{1}{\sqrt{n}}. \tag{8.11}$$

The rms condensation is therefore equal to the reciprocal of the square root of the number of particles in the system. For example, in a volume containing 10^{16} molecules (approximately 1 mm^3 under standard conditions),

$$\delta_{rms} = \frac{1}{\sqrt{10^{16}}} = 10^{-8}.$$

Since $\delta = \Delta V/V_0$, the rms fluctuation in volume (or density), in a volume of this size, is only one part in one hundred million. For smaller volumes the fluctuations increase and can contribute to the scattering of light, as has already been discussed.

From the expressions, the average potential energy of a system resulting from fluctuations in density can be calculated. From (8.8) the potential energy ϕ is given by the expression $\phi = n\kappa T/(2\delta^2)$. Hence the average potential energy, $\langle\phi\rangle$, equals the product of $n\kappa T/2$ and the mean-square condensation $\langle\delta^2\rangle$. From (8.11) we have $\langle\delta^2\rangle = 1/n$ and therefore $\langle\phi\rangle = \frac{1}{2}\kappa T$, i.e. the average potential energy of any volume element of an ideal gas due to thermal fluctuations is $\frac{1}{2}\kappa T$. This result is another illustration of the principle of equipartition. Note that ϕ depends on δ^2, not δ. The number of particles, n, does not appear in the expression for $\langle\phi\rangle$.

8.2   Fluctuations at a critical point

The phenomenon of scattering of light is especially pronounced in a substance at, or near, its critical point, producing a milky or opalescent appearance. This effect had been observed for many years but its cause was not known until the theory was formulated by Smoluchowski. For a substance at its critical point, we can evidently not assume that the equation of state of an ideal gas is obeyed, so we must return to the general series expansion for the pressure p given in (8.5). At the critical point

$$\frac{\partial p}{\partial V} = \frac{\partial^2 p}{\partial V^2} = 0$$

by definition, and the first nonvanishing derivative is $\partial^3 p/\partial V^3$. Keeping only this leading term in the series expansion in (8.5), the relation (8.4) becomes

$$\phi = -\int_{V_c}^{V}\left(\frac{\partial^3 p}{\partial V^3}\right)_c \frac{(V-V_c)^3}{3!}\, dV, \tag{8.12}$$

where V_0 was specified to be the critical volume V_c for the given critical temperature T_c.

To evaluate the derivative of the pressure we must know p as a function of V. To be specific, we assume that van der Waals' equation is obeyed. In that case

$$\left[p + a\left(\frac{n}{V}\right)^2\right](V - nb) = n\kappa T.$$

This result can be understood by noting that, by its definition, b is the volume that a mole of atoms or molecules can *not* occupy due to the repulsive forces at close packing; this is a consequence of there being a 'hard core' to atoms or molecules. Consequently nb has to be subtracted from the available volume V. The coefficient a is a measure of the attractive forces between slightly separated atoms or molecules, see also Fig. 8.1. The attraction between the

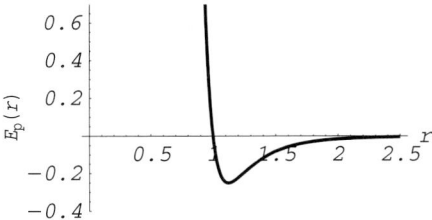

Figure 8.1 An illustration of the potential energy $E_p = A/r^{12} - B/r^6$ between two atoms at a separation r. This form, the Lennard–Jones potential, is often used for modeling van der Waals interactions. The model is a good approximation for monatomic gases, inert gases in particular. In the figure we have used $A = B = 1$.

individual particles constituting the gas means that, when an atom moves away from a high-density region, it is attracted back, slightly, toward the interior by the net attractive forces due to the excess of atoms there. This causes it to be slowed down a little, and consequently the selected atom does not hit the surrounding gas as hard as it does when these attractive forces are not present, such as, e.g., in a perfect gas (Jackson, 1968, Reichl, 1998). The consequence is a slight decrease in pressure compared with the case in which the weak attraction is not present. With a rearrangement of the foregoing expression, we find that

$$p = \frac{n\kappa T}{V - nb} - \frac{an^2}{V^2},$$

where the second term accounts for the decrease in pressure mentioned before. Using this relation, we find that $\partial p/\partial V = 0$ gives

$$2\frac{an^2}{V_c^3} = \frac{n\kappa T_c}{(V_c - nb)^2},$$

and that $\partial^2 p/\partial V^2 = 0$ gives

$$3\frac{an^2}{V_c^4} = \frac{n\kappa T_c}{(V_c - nb)^3},$$

implying that $V_c = 3nb$ and $T_c = (2/3)^3 a/b$.

From these results we find that

$$\left(\frac{\partial^3 p}{\partial V^3}\right)_c = -\frac{1}{81}\frac{a}{b^5}\left(\frac{1}{n}\right)^3 = -\frac{27}{8}\frac{\kappa T_c n}{V_c^4}. \tag{8.13}$$

Again introducing the condensation δ, we find, on evaluating (8.12), that

$$\frac{\phi}{\kappa T_c} = \left(\frac{3}{8}\right)^2 n\delta^4,$$

implying that

$$dN = \alpha \exp\left(-\frac{9n\delta^4}{64}\right) d\delta. \tag{8.14}$$

This expression is to be compared for instance with (8.9) obtained for an ideal gas. The fluctuations now depend on the fourth power of the condensation, rather than on the square.

The constant α is again determined by the requirement that $\int dN = N$. Using the definition of the gamma-function $\Gamma(n+1) = \int_0^\infty x^n \exp(-x)\, dx$, the result is

$$\frac{dN}{N} = \frac{\sqrt[4]{9n/64}}{2\Gamma(5/4)} \exp\left(-\frac{9n\,\delta^4}{64}\right) d\delta. \tag{8.15}$$

Again the mean condensation is zero, while the rms value, evaluated in the usual way, is

$$\delta_{\mathrm{rms}} = \frac{0.95}{\sqrt[4]{n}}, \tag{8.16}$$

where the numerical factor is approximate. Since the fluctuations now vary inversely with the fourth root of n, rather than with the square root, they can be significant even for relatively large values of n. For example, at the critical point of ethyl ether, in a volume of linear dimensions of the order of the wavelength of visible light, there are about 4×10^8 molecules, and the rms fluctuation in density is of the order of 1.5%. It is these relatively large density fluctuations which account for the opalescence observed at the critical point (Tolman, 1938, Lee et al., 1973).

9   A reference model

It is often a great advantage to have a simple model as a basis for discussion. Such a model should be based on some elementary statistical properties, and at least in principle allow relevant results to be obtained analytically. As an example we here consider a simple synthetic signal generated by a random superposition of a certain basic structure $\psi(t)$. In particular we shall use the model to illustrate the uncertainties associated with experimentally obtained average values, for instance when finite-length records are used. With the present model a time record becomes

$$\Phi(t) = \sum_{k=1}^{N} a\psi(t - t_k),\tag{9.1}$$

where the N positions labeled t_k are randomly distributed in a large interval $\mathcal{T}$. In the following t_k is denoted the 'pulse-arrival time' because t is most often a time-like variable, although this need not necessarily be the case. In most of the following analysis, the corresponding distribution is taken to be uniform, meaning that the probability $P(t_k) = 1/\mathcal{T}$ for $0 < t_k < \mathcal{T}$ and vanishing otherwise for all k. An amplitude a is included explicitly for convenience.

The time series (9.1) evidently constitutes a stationary random process and it will be demonstrated that it is ergodic by construction for large $\mathcal{T}$. Examples from actual realizations are shown in Fig. 9.1. The basic pulse was here taken to be $\psi(t) = \exp(-t^2)$ in all cases, with unit amplitude.

The model (9.1) was analyzed in great detail by Rice (1944, 1945, 1954) and also by Davenport and Root (1958). It is quite general, and can refer to many possible physically relevant situations, of which a few will be discussed later in this chapter.

9.1   Campbell's theorem

By considering a time record such as (9.1), a few basic results are readily obtained. Campbell's theorem (Campbell, 1909) states that the average signal is given by

$$\langle \Phi(t) \rangle = \mu a \int_{-\infty}^{\infty} \psi(t)\, dt,\tag{9.2}$$

and the mean-square value of the fluctuations about this average is

$$\langle (\Phi(t) - \langle \Phi(t) \rangle)^2 \rangle = \mu a^2 \int_{-\infty}^{\infty} \psi^2(t)\, dt,\tag{9.3}$$

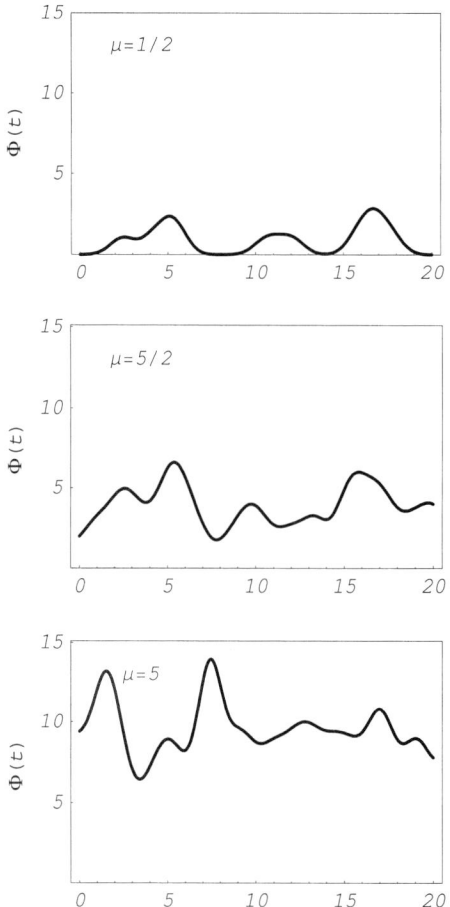

Figure 9.1 Examples of three realizations of the model for densities $\mu = N/\mathcal{T} = \frac{1}{2}, \frac{5}{2}$, and 5. Only a fraction of an extended sequence is shown.

where μ is the average number of basic structures per time unit. The statement of the theorem is not precise until what is meant by 'average' has been defined. From the form of the equations one might be tempted to think of a time average, e.g. the value

$$\lim_{\mathcal{T} \to \infty} \frac{1}{\mathcal{T}} \int_0^{\mathcal{T}} \Phi(t) \, dt. \tag{9.4}$$

However, in the proof of the theorem the average is generally taken over a great many intervals of length $\mathcal{T}$ with t held constant. This is the ensemble averaging discussed before. To make it clear in the present context we take the case of $\langle \Phi(t) \rangle$ for illustration. We observe $\Phi(t)$ for many, say M, intervals each of length $\mathcal{T}$, where $\mathcal{T}$ is large in comparison with the interval over which the effect $\psi(t)$ of a single basic structure, or pulse, is appreciable. Let $^n\Phi(t')$ be the value of $\Phi(t)$, t' seconds after the beginning of the nth interval. Thus, t' is equal to t plus a constant depending upon the time at which the interval begins. We place the subscript in front because we wish to reserve the usual place for another subscript later on. The value of $\langle \Phi(t') \rangle$ is then defined as

$$\langle \Phi(t') \rangle = \lim_{M \to \infty} \frac{1}{M} \left[{}^1\Phi(t') + {}^2\Phi(t') + \cdots + {}^M\Phi(t') \right].$$ (9.5)

This limit is assumed to exist. The mean-square value of the fluctuation of $\Phi(t')$ is defined in much the same way.

Actually, as equations (9.2) and (9.3) of Campbell's theorem show, these averages and all the similar averages encountered later turn out to be independent of the time. When this is true and when the M intervals in (9.5) are taken consecutively, the time average (9.4) and the average (9.5) become the same; more generally, the signal constitutes an ergodic random process. To show that the time average equals the ensemble average, we integrate both sides of (9.5) from 0 to T and obtain

$$\langle \Phi(t') \rangle = \lim_{M \to \infty} \frac{1}{MT} \sum_{m=1}^{M} \int_0^T {}^m\Phi(t') \, dt'$$

$$= \lim_{M \to \infty} \frac{1}{MT} \int_0^{MT} \Phi(t) \, dt.$$ (9.6)

This is the same as the time average (9.4), again assuming that the limit in (9.6) exists. It is evidently possible here also to choose the reference level so that $\langle \Phi(t) \rangle = 0$, but as the model's construction clearly indicates, this will not be a particularly smart thing to do!

9.1.1   Proof of Campbell's theorem

Consider the case in which exactly K pulses are present in the record of length T. We think of these K pulses as determined to arrive in the interval $\{0, T\}$ but any particular pulse is just as likely to arrive at one time as it is at any other time, i.e. the pulse-arrival times are uniformly distributed in the interval $\{0, T\}$. These pulses are numbered from 1 to K for purposes of identification, but it is to be emphasized that the numbering need not have anything to do with the order of arrival. Thus, if t_k be the time of arrival of pulse number k, the probability that t_k lies in the interval $\{t, t + dt\}$ is dt/T irrespective of the arrival of any other pulse.

We take T to be very large compared with the range of values of t for which $\psi(t)$ is appreciably different than zero. In physical applications such a range usually exists and we shall call it Δ even though it need not be very definite. Then, when exactly K pulses arrive in the interval $\{0, T\}$, the effect is given by (9.1), where we here include a subscript $_K$ to emphasize that the number of pulses is fixed in the given interval T. In other words, the intervals are *conditionally* selected with respect to K.

Suppose that we examine a large number M of intervals of length T. The number having exactly K arrivals will be, to a first approximation, $MP(K)$, where $P(K)$ is a Poisson distribution, see Appendix B. For a fixed value of t and for each interval having K arrivals, $\Phi_K(t)$ will have a definite value. As $M \to \infty$, the average value of the $\Phi_K(t)$'s, obtained by averaging over the intervals, is

$$\langle \Phi_K(t) \rangle = \int_0^T \frac{dt_1}{T} \cdots \int_0^T \frac{dt_K}{T} \sum_{k=1}^K a\psi(t - t_k)$$

$$= \sum_{k=1}^K \int_0^T \frac{dt_k}{T} \, a\psi(t - t_k). \tag{9.7}$$

If $\Delta < t < T - \Delta$, we have effectively

$$\langle \Phi_K(t) \rangle = \frac{K}{T} \int_{-\infty}^{\infty} a\psi(t) \, dt. \tag{9.8}$$

If we now average $\Phi(t)$ over all of the M intervals instead of only over those having K arrivals, we get by Bayes' rule, as $M \to \infty$,

$$\langle \Phi(t) \rangle = \sum_{K=0}^{\infty} \langle \Phi_K(t) \rangle P(K)$$

$$= \sum_{K=0}^{\infty} \frac{K}{T} \frac{(\mu T)^K}{K!} e^{-\mu T} \int_{-\infty}^{\infty} a\psi(t) \, dt$$

$$= \mu a \int_{-\infty}^{\infty} \psi(t) \, dt. \tag{9.9}$$

This proves the first part of Campbell's theorem. We have used this rather elaborate proof of the relatively simple result (9.9) in order to illustrate a method that may be used also for more complicated problems. Of course, (9.9) could be obtained by noting that the integral is the average value of the effect produced by the arrival of one pulse, the average being taken over one time unit, and that μ is the average number of arrivals per time unit.

In order to prove the second part, (9.3), of Campbell's theorem, we calculate first $\langle \Phi^2(t) \rangle$ and use

$$\langle (\Phi(t) - \langle \Phi(t) \rangle)^2 \rangle = \langle \Phi^2(t) \rangle - 2\langle \Phi(t) \langle \Phi(t) \rangle \rangle + \langle \Phi(t) \rangle^2$$

$$= \langle \Phi^2(t) \rangle - \langle \Phi(t) \rangle^2. \tag{9.10}$$

From the definition of $\Phi_K(t)$ we have

$$\Phi_K^2(t) = \sum_{k=1}^K \sum_{m=1}^K a^2 \psi(t - t_k)\psi(t - t_m).$$

Averaging this over all values of $t_1, t_2, \ldots, t_k$ with t held fixed as in (9.7), we have that

$$\langle \Phi_K^2(t) \rangle = \sum_{k=1}^K \sum_{m=1}^K \int_0^T \frac{dt_1}{T} \cdots \int_0^T \frac{dt_K}{T} a^2 \psi(t - t_k)\psi(t - t_m).$$

The multiple integral has two fundamentally different values. If $k = m$ its value is

$$\int_0^T a^2 \psi^2(t - t_k) \frac{dt_k}{T},$$

and if $k \neq m$ its value is

$$\int_0^T a\psi(t - t_k) \frac{dt_k}{T} \int_0^T a\psi(t - t_m) \frac{dt_m}{T}.$$

On counting the number of terms in the double sum, it is observed that, of the K^2 terms altogether, there are K having the first value and $K^2 - K$ having the second. Hence, if $\Delta < t < T - \Delta$, we have for large T

$$\langle \Phi_K^2(t) \rangle = \frac{K}{T} \int_{-\infty}^{\infty} a^2 \psi^2(t) \, dt + \frac{K(K-1)}{T^2} \left(\int_{-\infty}^{\infty} a \psi(t) dt \right)^2 .$$

The foregoing analysis was conditional, i.e. it considered only realizations containing exactly K pulses. By averaging over all the realizations where K is statistically distributed we find, using that $\langle K(K-1) \rangle = \langle K \rangle^2$ for a Poisson distribution for K, that

$$\langle \Phi^2(t) \rangle = \sum_{K=0}^{\infty} P(K) \langle \Phi_K^2(t) \rangle$$

$$= \mu a^2 \int_{-\infty}^{\infty} \psi^2(t) \, dt + \langle \Phi(t) \rangle^2,$$

where the summation with respect to K is performed as in (9.9), and after summation the value (9.9) for $\langle \Phi(t) \rangle$ is used. Comparison with (9.10) establishes the second part of Campbell's theorem. The coefficient μ in front of the integral is obtained even when the distribution of K differs from a Poisson distribution.

9.1.2   Extension of Campbell's theorem

Consider now a slight generalization of the signal (9.1) in the form

$$\Phi(t) = \sum_{k}^{N} a_k \psi(t - t_k), \tag{9.11}$$

where $\psi(t)$ is an *a priori* given function as before, and where $a_1, \ldots, a_k, \ldots$ are independent random variables all having the same distribution. It is assumed that all of the moments $\langle a^n \rangle$ exist, and that the events occur at random with a uniform distribution of t_k as before.

The extension states that the nth semi-invariant of the probability density $P(\Phi)$ of Φ, where Φ is given by (9.11), is

$$\lambda_n = \mu \langle a^n \rangle \int_{-\infty}^{\infty} [\psi(t)]^n \, dt, \tag{9.12}$$

where μ is the expected number of events per second. The semi-invariants (or *cumulants*) of a distribution are defined as the coefficients in the expansion

$$\ln \langle e^{i \Phi u} \rangle = \sum_{n=1}^{N} \frac{\lambda_n}{n!} (iu)^n + \mathcal{O}(u^N), \tag{9.13}$$

i.e. as the coefficients in the expansion of the logarithm of the characteristic function. The λ's are related to the moments of the distribution. Thus, if $m_1, m_2, \ldots$ denote the first, second, $\ldots$ moments about zero, we have

$$\langle e^{i \Phi u} \rangle = 1 + \sum_{n=1}^{N} \frac{m_n}{n!} (iu)^n + \mathcal{O}(u^N).$$

By combining this relation with the one defining the λ's, it may be shown that

$$\langle \Phi \rangle = m_1 = \lambda_1,$$
$$\langle \Phi^2 \rangle = m_2 = \lambda_2 + \lambda_1 m_1,$$
$$\langle \Phi^3 \rangle = m_3 = \lambda_3 + 2\lambda_2 m_1 + \lambda_1 m_2.$$

It follows that $\lambda_1 = \langle \Phi \rangle$ and $\lambda_2 = \langle (\Phi - \langle \Phi \rangle)^2 \rangle$. Hence (9.12) yields the original statement of Campbell's theorem when we set n equal to unity.

- **Exercise:** Consider a process that is the sum of several independent processes. Demonstrate that the cumulants, or semi-invariants, associated with this process are given by the sum of cumulants for the individual processes.

- **Exercise:** Demonstrate the following relations for the cumulants λ_n in terms of the averages $m_n \equiv \langle \Phi^n \rangle$:

$$\lambda_1 = m_1,$$
$$\lambda_2 = m_2 - m_1^2 \equiv \sigma^2,$$
$$\lambda_3 = m_3 - 3m_1 m_2 + 2m_1^3,$$
$$\lambda_4 = m_4 - 4m_1 m_3 - 3m_2^2 + 12 m_2 m_1^2 - 6m_1^4.$$

9.1.2.1   Further extensions of Campbell's theorem

Other extensions of Campbell's theorem may be made. The foregoing analysis might give the impression that the uniform distribution of pulse-arrival times t_k was essential. The following example, given by Rice (1944, 1945, 1954), demonstrates that the analysis can be extended to more general pulse-arrival-time distributions. For example, suppose that, in the expression (9.11) for $\Phi(t)$, $t_1, t_2, \ldots, t_k, \ldots$, while still random variables, are no longer uniformly distributed as was assumed before. Suppose instead that a probability density $P(x)$ is given, such that $x_j > 0$ is the interval between two successive events labeled by j and $j + 1$:

$$t_2 = t_1 + x_1,$$
$$t_3 = t_2 + x_2 = t_1 + x_1 + x_2. \tag{9.14}$$
$$\vdots$$

We can use the arguments from Section 3.2 and show that, for the case treated before,

$$P(x) = \mu e^{-\mu x}. \tag{9.15}$$

See also Appendix B. For other distributions more regular separations of pulses can be modeled. We can for instance illustrate a certain 'recovery time' for the pulse emission. Pulses appearing with fixed intervals is an extreme case. It is assumed in the following that the expected number of events per time unit is still μ.

Consider again the series (9.11). For a long time interval extending from $t = t_1$ to $t = T + t_1$ inside of which there are exactly K events, we have by writing out the individual terms in (9.11) for clarity

$$\Phi(t) = a_1 \psi(t - t_1) + a_2 \psi(t - t_1 - x_1) + \cdots + a_{K+1} \psi(t - t_1 - x_1 \cdots - x_K)$$
$$= a_1 \psi(t') + a_2 \psi(t' - x_1) + \cdots + a_{K+1} \psi(t' - x_1 - \cdots - x_K),$$

$$\Phi^2(t) = a_1^2 \psi^2(t') + a_2^2 \psi^2(t' - x_1) + \cdots + a_{K+1}^2 \psi^2(t' - x_1 - \cdots - x_K)$$
$$+ 2a_1 a_2 \psi(t')\psi(t' - x_1) + \cdots$$
$$+ 2a_1 a_{K+1} \psi(t')\psi(t' - x_1 - \cdots - x_K)$$
$$+ 2a_2 a_3 \psi(t' - x_1)\psi(t' - x_1 - x_2) + \cdots + \cdots,$$

where $t' = t - t_1$. If we integrate $\Phi^2(t)$ over the entire interval $0 < t' < T$ and drop the primes, we get approximately

$$\int_0^T \Phi^2(t)\,dt = (a_1^2 + \cdots + a_{K+1}^2)\varphi(0)$$
$$+ 2a_1 a_2 \varphi(x_1) + 2a_1 a_3 \varphi(x_1 + x_2) + \cdots$$
$$+ 2a_1 a_{K+1} \varphi(x_1 + \cdots + x_k)$$
$$+ 2a_2 a_3 \varphi(x_2) + \cdots + 2a_K a_{K+1} \varphi(x_K), \qquad (9.16)$$

where we have defined the convolution

$$\varphi(x) = \int_{-\infty}^{\infty} \psi(t)\psi(t - x)\,dt. \qquad (9.17)$$

Now divide both sides of (9.16) by T and consider K and T to be very large. Then the following individual terms enter the expression:

$$\frac{K}{T}\frac{a_1^2 + \cdots + a_{K+1}^2}{K}\varphi(0) \approx \mu\varphi(0)\langle a^2\rangle,$$

$$\frac{1}{T}\left[a_1 a_2 \varphi(x_1) + a_2 a_3 \varphi(x_2) + \cdots + a_K a_{K+1}\varphi(x_K)\right] = \frac{K}{T}\langle a_k a_{k+1}\varphi(x_k)\rangle$$
$$\approx \mu\langle a\rangle^2 \int_0^{\infty} \varphi(x)P(x)\,dx,$$

and also

$$\frac{1}{T}\left[a_1 a_3 \varphi(x_1 + x_2) + \cdots\right] = \frac{K-1}{T}\langle a_k a_{k+2}\varphi(x_k + x_{k+1})\rangle$$
$$\approx \mu\langle a\rangle^2 \int_0^{\infty} dx_1 \int_0^{\infty} dx_2\, P(x_1)P(x_2)\varphi(x_1 + x_2).$$

Consequently (9.16) can be reduced to

$$\langle\Phi^2(t)\rangle = \lim_{T\to\infty}\frac{1}{T}\int_0^T \Phi^2(t)\,dt$$
$$= \mu\langle a^2\rangle\varphi(0) + 2\mu\langle a\rangle^2\left(\int_0^{\infty} P(x)\varphi(x)\,dx\right.$$
$$\left. + \int_0^{\infty} dx_1 \int_0^{\infty} dx_2\, P(x_1)P(x_2)\varphi(x_1 + x_2) + \cdots\right).$$

The foregoing results are rather general. We consider here the special, but important, case for which

$$\psi(t) = 0, \qquad t < 0,$$
$$\psi(t) = e^{-\alpha t}, \qquad t > 0. \qquad (9.18)$$

For this particular exponential form for $\psi(t)$ we obtain the convolution (9.17) as

$$\varphi(x) = \frac{e^{-\alpha x}}{2\alpha}.$$

The multiple integrals appearing in the expression for $\langle \Phi^2(t) \rangle$ may be written in terms of powers of the integral:

$$q \equiv \int_0^\infty P(x)e^{-\alpha x}\, dx. \tag{9.19}$$

Thus

$$2\alpha\langle\Phi^2(t)\rangle = \mu\langle a^2\rangle + \langle a\rangle^2 \frac{2\mu q}{1-q},$$

and, since $\langle\Phi(t)\rangle = \mu\langle a\rangle \int_{-\infty}^\infty \psi(t)\, dt = \langle a\rangle\mu\alpha$, we have

$$\langle\Phi^2(t)\rangle - \langle\Phi(t)\rangle^2 = \frac{\mu\langle a^2\rangle}{2\alpha} + \left(\frac{\mu\langle a\rangle}{\alpha}\right)^2 \left(\frac{\alpha q}{\mu(1-q)} - 1\right). \tag{9.20}$$

Equations (9.19) and (9.20) give an extension of Campbell's theorem subject to the restrictions discussed in connection with equations (9.14) and (9.18). Other generalizations have been made by Rowland (1936).

- **Exercise:** Verify that (9.20) gives the correct answer when $P(x)$ is given by (9.15), and investigate also the case in which the events are spaced equally.

9.2   Probability densities

The foregoing results for $\langle\Phi(t)\rangle$ and $\langle\Phi^2(t)\rangle$ could, as has been demonstrated, be obtained without the relevant probabilities being explicitly known. The one-point amplitude probability density for the model can, however, also be obtained by use of characteristic functions (Rice, 1944, 1945, 1954). The problem is evidently that of obtaining $P(\Phi)$ on the basis of the *a priori* given probability density of t_k, i.e. $P(t_k)$.

The probability that a sum of K independent variables $x_1 + x_2 + \cdots + x_K$ lies in the range $[X, X + dX]$ is

$$dX \frac{1}{2\pi} \int_{-\infty}^\infty \exp(-iXu) \prod_{l=1}^K \langle\exp(ix_k u)\rangle\, du,$$

i.e. the characteristic function for a sum of independent variables is the product of their individual characteristic functions. These functions, $\langle\exp(ix_k u)\rangle$, associated with the distribution of x_k are obtained by averaging over the values of x_k. By identifying x_k with $a\psi(t - t_k)$ we find that

$$\langle\exp[ia\psi(t - t_k)u]\rangle = \frac{1}{T} \int_0^T \exp[iua\psi(t - t_k)]\, dt_k,$$

noting that the time t_k is here the only statistically varying quantity. All of the K characteristic functions are the same, so

$$P_K(\Phi) = \frac{1}{2\pi} \int_{-\infty}^{\infty} \exp(-i\Phi u) \left(\frac{1}{T} \int_0^T \exp[iua\psi(t-\tau)]\, d\tau \right)^K du.$$

This expression gives a *conditional* probability assuming K to be given. The averaging over all K is performed assuming that the probability density is $P(K) = e^{-\mu T}(\mu T)^K/K!$, i.e. a Poisson distribution for K. With $P(\Phi) = \sum_{K=0}^{\infty} P(K)P_K(\Phi)$ we perform the summation by use of $e^{\mu S} = \sum_{K=0}^{\infty}(\mu S)^K/K!$ to obtain after some elementary manipulations,

$$P(\Phi) = \frac{1}{2\pi} \int_{-\infty}^{\infty} \exp\left(-i\Phi u + \mu \int_{-\infty}^{\infty} \{\exp[iua\psi(t)] - 1\}\, dt \right) du, \tag{9.21}$$

where the density of structures $\mu = \langle N \rangle / T$ was introduced in terms of the average number, $\langle N \rangle$, of structures in a record. Note that $P(\Phi)$ is real as it should be, since $P(\Phi) = P^*(\Phi)$.

The two-time amplitude, or joint probability density, and also higher order densities can be obtained similarly. For instance the two-time probability density becomes

$$P(\Phi, \Psi) = \frac{1}{(2\pi)^2} \int \int_{-\infty}^{\infty} \exp(-i\Phi u - i\Psi v)$$

$$\times \exp\left(\mu \int_{-\infty}^{\infty} \{\exp[iua\phi(t) + ivb\psi(t)] - 1\}\, dt \right) du\, dv, \tag{9.22}$$

where $\Phi = \Phi(t)$ and $\Psi = \Psi(t)$ in general correspond to two records. For the present application we have $\Phi(t) = \sum_k^N a\psi(t - t_k)$ and $\Psi(t) \equiv \Phi(t + \tau) = \sum_k^N a\psi(t - t_k + \tau)$.

- **Exercise:** Demonstrate that the result for the probability density becomes particularly simple when there are two pulse types having the same density and the same form but with opposite polarities.

- **Exercise:** Consider the particularly simple case in which the basic pulse or structure is composed of one or more boxes as illustrated in Fig. 9.2. Demonstrate that the probability density for case I is

$$P(\Phi) = \frac{\mu^N}{N!} e^{-\mu} \delta(\Phi - N), \qquad \text{for } N = 1, 2, \ldots,$$

where μ is the density of pulses. Demonstrate that, for case II as well as case III in Fig. 9.2, the probability density, in terms of the modified Bessel functions I_N of order N, is

$$P(\Phi) = I_N(\mu) e^{-\mu} \delta(\Phi - N), \qquad \text{for } N = 1, 2, \ldots,$$

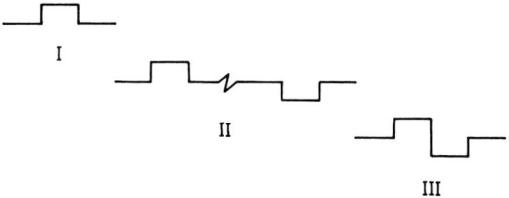

Figure 9.2 An illustration of particularly simple basic structures or pulses composed of one or more boxes or bins of unit amplitude.

i.e. an even distribution with all odd semi-invariants vanishing, while all even semi-invariants are equal to μ. The *two-point* probabilities for cases II and III are evidently different but the one-point probability densities, or *projections*, are the same.

9.2.1   The extended model

An extension of Campbell's theorem for distributed values of a follows almost at once from the generalization of expression (9.21) for the probability density $P(\Phi)$. By proceeding as in the foregoing section and identifying x_k with $a_k \psi(t - t_k)$, we first find that

$$\langle \exp(ix_k u) \rangle = \frac{1}{T} \int_{-\infty}^{\infty} q(a) \int_0^T \exp[iua\psi(t - t_k)] \, dt_k \, da,$$

where $q(a)$ is the probability density function for the a's. It is readily demonstrated that the probability density $P(\Phi)$ of Φ defined by (9.11) is

$$P(\Phi) = \frac{1}{2\pi} \int_{-\infty}^{\infty} \exp\left(-i\Phi u + \mu \int_{-\infty}^{\infty} q(a) \, da \int_{-\infty}^{\infty} [\exp(iua\psi(t)) - 1] \, dt \right) du. \qquad (9.23)$$

The logarithm of the characteristic function of $P(\Phi)$ is, from (9.23),

$$\mu \int_{-\infty}^{\infty} q(a) \, da \int_{-\infty}^{\infty} [\exp(iua\psi(t)) - 1] \, dt$$

$$= \sum_{n=1}^{\infty} \frac{(iu)^n}{n!} \mu \int_{-\infty}^{\infty} q(a) \, da \, a^n \int_{-\infty}^{\infty} \psi^n(t) \, dt.$$

Comparison with the series (9.13) defining the semi-invariants gives the extension of Campbell's theorem stated by (9.12).

The analysis given hitherto assumed that there was one basic pulse shape only. It is straightforward to generalize the results to the case in which many different pulses independently enter the construction of a signal. Thus, for instance, (9.21) is generalized to

$$P(\Phi) = \frac{1}{2\pi} \int_{-\infty}^{\infty} \exp\left(-i\Phi u + \sum_n \mu_n \int_{-\infty}^{\infty} \{\exp[iu\psi_n(t)] - 1\} \, dt \right) du, \qquad (9.24)$$

where the summation runs over all different pulse shapes ψ_n with densities μ_n. The result (9.24) is easily verified by recalling that the characteristic function for a sum of independent variables is the product of their respective, individual, characteristic functions.

9.2.2   Transition to Gaussian distributions

It can be demonstrated analytically that $P(\Phi)$ approaches a Gaussian distribution when $\mu \to \infty$, as expected with reference to the central-limit theorem, since in this limit many pulses contribute to the signal at any time. To demonstrate this transition (Rice, 1954), it is advantageous to express $P(\Phi)$ as

$$P(\Phi) = \frac{1}{2\pi} \int_{-\infty}^{\infty} \exp\left(-i\Phi u + \sum_{n=1}^{\infty} (iu)^n \lambda_n/n!\right) du, \tag{9.25}$$

using $\exp(x) = \sum_{n=0}^{\infty} x^n/n!$, where the semi-invariants, λ_n, are defined as before by

$$\lambda_n = \mu \int_{-\infty}^{\infty} [a\psi(t)]^n \, dt. \tag{9.26}$$

Setting $\lambda_2 \equiv \sigma^2$ and introducing the new variable $x = (\Phi - \lambda_1)/\sigma$, the exponential $\exp \sum_{n=1}^{\infty} (iu)^n \lambda_n/n!$ is expanded as a power series in u. After termwise integration the terms are collected according to their order in powers of $\mu^{-1/2}$, giving

$$P(\Phi) \sim f^{(0)}(x)/\sigma - \frac{\lambda_3}{3!\sigma^4} f^{(3)}(x) + \left[\frac{\lambda_4}{4!\sigma^5} f^{(4)}(x) + \frac{\lambda_3^2}{72\sigma^7} f^{(6)}(x)\right] + \cdots. \tag{9.27}$$

The function $f(x)$ is introduced through

$$\frac{1}{2\pi} \int_{-\infty}^{\infty} (iu\sigma)^n \exp(-iu\sigma x - u^2\sigma^2/2) \, du = (-1)^n f^{(n)}(x)/\sigma, \tag{9.28}$$

where

$$f^{(n)}(x) = \frac{1}{\sqrt{2\pi}} \frac{d^n}{dx^n} e^{-x^2/2}. \tag{9.29}$$

Since λ_n scales as μ^n we find that the first term in (9.27) is $\mathcal{O}(\mu^{-1/2})$ while the second term is $\mathcal{O}(\mu^{-1})$, and the term within the brackets $\mathcal{O}(\mu^{-3/2})$, etc. This is called the Edgeworth series. The first term gives the normal distribution and the remaining terms show how it is approached as $\mu \to \infty$. The proof given here is based on a somewhat restrictive model and will not serve as a general proof of the central-limit theorem, although it is closely related to it.

The transition to multidimensional Gaussian, or normal, distributions for $\mu \to \infty$ is demonstrated similarly for multipoint distributions. In this Gaussian limit it can be argued that all the statistical information is contained in the correlation function.

- **Exercise:** As has already been mentioned, the transition to Gaussian statistics can be performed by keeping the mean-square level of fluctuation constant, i.e., $\mu a^2 = \text{constant}$, while $\mu \to \infty$, implying that $a \to 0$. Demonstrate that the triple-correlation function vanishes in this limit as $1/\sqrt{\mu}$, provided that $\int_{-\infty}^{\infty} \psi(t) \, dt = 0$. How about higher order correlations?

9.2.2.1   An apparent paradox

There remains a paradox in the foregoing analysis for the case in which the basic pulse in the time series takes discrete values only as in Fig. 9.2, i.e. the probability is nonzero for discrete values only. The question is that of how such a discrete distribution can be approximated by a continuous Gaussian probability distribution in the limit $\mu \to \infty$. To study this particular question explicitly, we consider a stochastic variable, X, that takes the values 0 and 1 with probability $\frac{1}{2}$ each. Let Y be the sum of r such variables. Then Y takes the values $n = 0, 1, 2, 3, \ldots, r$ with probability

$$P_n = 2^{-r} \binom{r}{n},$$

see Appendix A. On noting the form of the average $\langle Y \rangle = r\langle X \rangle = \frac{1}{2}r$ and variance $\sigma_Y^2 = r\sigma_X^2 = \frac{1}{4}r$, it is found advantageous (van Kampen, 1981) to introduce a new variable Z, given through $Y = \frac{1}{2}r + \frac{1}{2}Z\sqrt{r}$. The probability that Z lies in the interval $(z; z + \Delta z)$ is

$$P_Z(z)\,\Delta z = \sum_{\frac{1}{2}r + \frac{1}{2}z\sqrt{r} < n < \frac{1}{2}r + \frac{1}{2}(z+\Delta z)\sqrt{r}} P_n.$$

For large r we have

$$\log P_n = -r\log 2 + (r + \tfrac{1}{2})\log r - (n + \tfrac{1}{2})\log n$$
$$- (r - n + \tfrac{1}{2})\log(r - n) - \tfrac{1}{2}\log(2\pi)$$
$$= \log 2 - \tfrac{1}{2}\log r - (\tfrac{1}{2}r + \tfrac{1}{2}r^{1/2}z + \tfrac{1}{2})(r^{-1/2}z - \tfrac{1}{2}r^{-1}z^2 + \cdots)$$
$$- (\tfrac{1}{2}r - \tfrac{1}{2}r^{1/2}z + \tfrac{1}{2})(-r^{-1/2}z - \tfrac{1}{2}r^{-1}z^2 + \cdots) - \tfrac{1}{2}\log(2\pi)$$
$$= \log 2 - \tfrac{1}{2}\log r - \tfrac{1}{2}z^2 - \tfrac{1}{2}\log(2\pi) + \mathcal{O}(r^{-1/2}).$$

Consequently

$$P_Z(z)\,\Delta z = \tfrac{1}{2}r^{1/2}\,\Delta z\,\exp[\log 2 - \tfrac{1}{2}\log r - \tfrac{1}{2}z^2 - \log(2\pi)]$$
$$= \frac{1}{\sqrt{2\pi}}e^{-z^2/2}\,\Delta z. \tag{9.30}$$

The implication of this result is that, for large r, we can use the approximation

$$P_Y(y) = \sqrt{\frac{2}{\pi r}}\exp\left(-\frac{(y - \frac{1}{2}r)^2}{\frac{1}{2}r}\right). \tag{9.31}$$

This is evidently a coarse-grained distribution; it gives the probability of finding Y in an interval $y, y + \Delta y$ when $\Delta y \gg 1$. It is obviously incorrect to use it when $\Delta y \leq 1$. It may also seem paradoxical that the range of (9.30) extends from $-\infty$ to ∞ whereas Y by construction can assume positive values only. The normalization $\int_{-\infty}^{\infty} P_Y(y)\,dy = 1$ of the probability density will be violated if the range is reduced to $\{0, \infty\}$. Again it is emphasized that the distribution is only an approximation and, when r is large, it is actually exceedingly close to zero for all $y < 0$.

9.3   Correlation functions

The two-point correlation function associated with the model (9.1) can be calculated directly without use of the two-point probability density. First select all records with K events and calculate

$$R_K(\tau) \equiv \langle \Phi_K(t)\Phi_K(t+\tau)\rangle$$
$$= \sum_{k=1}^{K}\sum_{m=1}^{K}\int_0^T \frac{dt_1}{T}\cdots\int_0^T \frac{dt_K}{T}\,a^2\psi(t - t_k)\psi(t + \tau - t_m).$$

This expression contains two different types of terms: those in which $k = m$, there are K of these; and then $K^2 - K$ terms in which $k \neq m$. For the latter type t_k and t_m are independent and the average of the product becomes a product of averages. The result is

$$R_K(\tau) = \frac{K}{T} a^2 \int_{-\infty}^{\infty} \psi(t)\psi(t + \tau) \, dt + \frac{K(K-1)}{T^2} \left(a \int_{-\infty}^{\infty} \psi(t) dt \right)^2,$$

where it was assumed that T is so large that it is safe to let the integration limits go to infinity. Assuming, as before, that we have a Poisson distribution for K, we have $\langle K(K-1)\rangle = \langle K \rangle^2$. With $\mu = \langle K \rangle / T$ the final result becomes

$$\mathcal{R}(\tau) = \langle \Phi(t)\Phi(\tau + t) \rangle$$

$$= \mu a^2 \int_{-\infty}^{\infty} \psi(t)\psi(\tau + t) \, dt + \left(\mu a \int_{-\infty}^{\infty} \psi(t) \, dt \right)^2. \tag{9.32}$$

In particular, the mean-square level of fluctuation is obtained from $\mathcal{R}(0)$ when the dc level is subtracted. It may be worth noting that the assumption of a Poisson distribution for K was used explicitly in the second term of (9.32) only. This second term vanishes for many cases of practical interest. In these cases $\mathcal{R}(t)$ is unaffected by the transition to the Gaussian limit for $\mu \to \infty$, provided that $a \to 0$ so that the mean-square level of fluctuation of the signal μa^2 is kept constant.

   The correlation function, and thus also the power spectrum, is insensitive to the sign of the basic structure, as well as to a 'mirroring' (i.e. replacing) of $\psi(t)$ by $\psi(-t)$. The power spectrum can not distinguish the direction of time, and remains the same if you read the time series (9.1) 'backward.' Note that *any* prescribed acceptable power spectrum (and by the Wiener–Khinchine theorem therefore any correlation function) can be modeled by proper choice of $\psi(t)$, but, given a correlation function such as (9.32) alone, it is not possible uniquely to reconstruct the basic pulse which was used. The correlation function (9.32) is related to the function defined in (9.17), but different notations are used because the two functions have different meanings.

- **Exercise:** Obtain the analytic expression for the correlation function associated with a record generated by a random superposition of the pulses shown in Fig. 9.2.

- **Exercise:** Electric fields are often detected by measuring the potential difference between two spatially separated probes, and then dividing the signal by the probes' separation to get a result in units of $V\,m^{-1}$. (Electric fields in the ionosphere are often determined this way by use of instrumented rockets, for instance.) When the scale length of the variation in electric field is much larger than the interprobe separation, this method works well. Illustrate the limitations of the procedure by generating a spatial variation in potential by a random superposition of many identical pulses with a known shape. Sample the difference signal at two positions with fixed separation, but let the 'probe pair' move with constant speed to generate a temporally varying signal as in (9.1). Obtain an analytic result for the correlation function and the corresponding power spectrum for the difference signal, and illustrate the transition when the width of the basic pulse is first much *larger* than the interprobe separation, and eventually becomes much *smaller*.

- **Exercise:** Obtain the analytic expression for the triple-correlation function

$$S(\tau_1, \tau_2) = \langle \Phi(t)\Phi(t + \tau_1)\Phi(t + \tau_2) \rangle$$

by a similar analysis for the model (9.1). Unlike the correlation function, the triple correlation changes sign with $\psi(t)$.

- **Exercise:** Assume that it is given *a priori* that an ensemble of time series is generated by the model described by (9.1) with only one basic structure $\psi(t)$ entering the construction. Demonstrate that $\psi(t)$ can be uniquely determined if the correlation function, $\mathcal{R}(\tau)$, and the triple-correlation function, $S(\tau_1, \tau_2)$, are known. Can you, by proper choice of the basic structure $\psi(t)$ in (9.1), generate an ensemble of time series having *any* physically acceptable prescribed set of correlation and triple-correlation functions?

- **Exercise:** In this chapter all signals were assumed to be continuous functions of the independent variable, the time t in particular. In reality, the signal is often sampled, and available only at discrete times. Most of the analysis is readily generalized to discretized representations. There is a wealth of literature on the subject.

   Demonstrate that a nonperiodic signal that is sampled in time at regular intervals, Δt, has a power spectrum that is continuous, with a periodicity $f_s = 1/\Delta t$, f_s being the *sampling frequency*. Demonstrate that a periodic signal, with period $\mathcal{T}$, which is sampled in time at regular intervals Δt, has a power spectrum that is periodic with $f_s = 1/\Delta t$, and discrete with a frequency-sampling interval of $1/\mathcal{T}$. It is implicit that $\mathcal{T} = N\Delta t$, with N an integer.

An apparent paradox is concerned with the possible cases in which the correlation function assumes a local zero value at a certain time delay, say τ_*. This can happen when the basic pulse in (9.1) assumes positive as well as negative values, see (9.32). This will remain so also in the Gaussian limit where the density of pulses goes to infinity. Now this can appear puzzling; two uncorrelated events are also independent in the Gaussian limit. This means that, for correlations of the particular sort mentioned, here we start out for small time delays having statistical *dependence*. For a certain time delay, τ_*, the correlation vanishes and the events are statistically independent. Then, for time delays larger than τ_*, they become dependent again! This is, however, not really a paradox, insofar as, for any finite pulse density μ, no matter how large, there will be small terms of importance in the series expansion in the exponents in the expression for the probability density also at the time delay for which the correlation vanishes. These terms will ensure that the events with time separation τ_* retain their statistical dependence.

9.3.1   Cross-correlation functions

Assume that two different events, described by their respective time series $\Phi(t)$ and $\Psi(t)$, have a common cause. A certain random event at time t_k is thus assumed to excite pulses $\phi(t - t_k)$ and $\psi(t - t_k)$ in two different records. By a straightforward generalization of the foregoing analysis it is readily demonstrated that the correlation between the two records $\Phi(t)$ and $\Psi(t)$ can be obtained as

$$
\begin{aligned}
\mathcal{R}_{\Phi\Psi} &= \langle \Phi(t)\Psi(t) \rangle \\
&= \mu \int_{-\infty}^{\infty} \phi(t)\psi(t)\,dt + \left(\mu \int_{-\infty}^{\infty} \phi(t)\,dt \right)\left(\mu \int_{-\infty}^{\infty} \psi(t)\,dt \right),
\end{aligned} \tag{9.33}
$$

where the density, μ, of structures ϕ and ψ is the same in both records by construction. The result (9.33) is due to Rowland (1936). It contains in particular (9.32) as a special case since it can be argued that the relation must be valid also if we identify $\psi(t)$ with $\phi(t+\tau)$. The result (9.33) can be generalized to other cases as well. Note that $\int \psi(t)\,dt = 0$ and $\int \phi(t)\,dt = 0$ for *orthogonal pulses*, $\int \phi(t)\psi(t)\,dt = 0$, implies that $\mathcal{R}_{\Phi\Psi} = 0$, i.e. the events are uncorrelated, but quite evidently not *independent*, since they are placed synchronously in the two records.

Again we can delay one time record with respect to the other and obtain the straightforward and almost trivial generalization of (9.33)

$$
\mathcal{R}_{\Phi\Psi}(\tau) = \mu \int_{-\infty}^{\infty} \phi(t)\psi(t+\tau)\,dt + \left(\mu \int_{-\infty}^{\infty} \phi(t)\,dt \right)\left(\mu \int_{-\infty}^{\infty} \psi(t)\,dt \right).
$$

Even if the basic pulses are orthogonal, $\mathcal{R}_{\Phi\Psi}(0) = 0$, we generally have $\mathcal{R}_{\Phi\Psi}(\tau) \neq 0$ for *some* values of τ. At first sight this seems surprising, since the statement apparently holds also for statistically independent records. Note, however, that the two records discussed here are by construction related, as has already been mentioned.

- **Exercise:** Consider a one-dimensional system, such that pulses propagate without change in shape along the x-axis, with statistically distributed velocities, with a given probability density for the velocities. Determine the cross-correlation of the two signals obtained by sampling the system at two positions with known separation.

9.4   Spectral representation

An alternative presentation of the signal (9.1) can be given in the form

$$
\Phi(t) = \frac{a_0}{2} + \sum_{n=1}^{K} \left[a_n \cos\left(\frac{2\pi n t}{T} \right) + b_n \sin\left(\frac{2\pi n t}{T} \right) \right]. \tag{9.34}
$$

The correspondence to (9.1) is obtained by the Fourier expansion of (9.1) over the interval T, for a given N:

$$
\begin{aligned}
a_n - ib_n &= \frac{2}{T} \sum_{j=1}^{N} \int_0^T a\psi(t-t_j) e^{-i2\pi n t/T}\,dt \\
&= \left(\frac{2}{T} \int_{-\infty}^{\infty} a\psi(t) e^{-i2\pi n t/T}\,dt \right) \sum_{j=1}^{N} e^{-in\theta_j},
\end{aligned} \tag{9.35}
$$

with $\theta_j = 2\pi t_j/T$. Since t_j for $j = 1, 2, \ldots, N$ is uniformly distributed over T by assumption, the θ_j's are random variables distributed uniformly over the interval $\{0, 2\pi\}$. The summation in

(9.35) can be interpreted as a sum over N randomly oriented vectors with a length given by the term in large brackets in (9.35). When N becomes large, which it does when the density $\mu \to \infty$ with T fixed, the real and imaginary parts of this sum are random variables, which tend to become independent and normally distributed about zero (Rice, 1944, 1945, 1954, Papoulis, 1991).

For any pulse density μ it can be shown that

$$\langle a_n \rangle = \langle b_n \rangle = 0, \qquad n \neq 0, \tag{9.36}$$

$$\langle a_n^2 \rangle = \langle b_n^2 \rangle = \sigma_n^2 \equiv \frac{2\mu}{T} \left| \int_{-\infty}^{\infty} a\psi(t) e^{-i2\pi n t/T} dt \right|^2, \tag{9.37}$$

$$\langle a_n b_m \rangle = \langle a_n a_m \rangle = \langle b_n b_m \rangle = 0, \qquad n \neq m. \tag{9.38}$$

The integration limits in (9.37) as well as (9.35) can be set to infinity when T is much larger than any timescale characterizing the individual pulse $\psi(t)$. The first one of these results (9.36) is trivially obtained and the other two follow by some simple algebra from (9.35). The expression (9.37) for the Fourier coefficients is consistent with the Wiener–Khinchine theorem expressing the power spectrum associated with the ensemble of records given in terms of the Fourier transform of the correlation function (9.29). The results (9.38) show that the coefficients corresponding to different Fourier components are uncorrelated; for Gaussian signals they are consequently also statistically independent. This observation is usually expressed as the statement that, for Gaussian signals, the phases of spectral components are uniformly distributed and statistically independent.

The power spectrum obtained by Fourier transformation of the correlation function is not sensitive to internal phase relations among the spectral components in the individual pulses. The bispectrum obtained by Fourier transformation of the triple-correlation function will reveal these phase relations and is therefore of considerable interest for analysis of data (Hasselman *et al.*, 1963, Lii *et al.*, 1976, Kim and Powers, 1979). Being able to deal with normalized auto-correlations and triple correlations is often an advantage. The auto-correlation is readily normalized by its value, σ^2, at zero time delay. This is not quite so easy for the triple correlation, which may very well be vanishing at the origin. An appropriate normalization is for instance

$$\langle \Phi(t)\Phi(t + \tau_1)\Phi(t + \tau_2) \rangle \Big/ \sqrt{\langle \Phi^2(t)\Phi^2(t + \tau_1) \rangle \langle \Phi^2(t) \rangle},$$

with the result confined to the interval $\{-1, 1\}$ by the Schwartz inequality. (Other possible normalizations need not have this property.) The normalizing quantities can be given analytic expressions by use of higher order correlation functions such as

$$\mathcal{Q}(\tau_1, \tau_2, \tau_3) = \langle \Phi(t)\Phi(t + \tau_1)\Phi(t + \tau_2)\Phi(t + \tau_3) \rangle,$$

and the correlation $\mathcal{R}(\tau)$. Also the normalized triple-correlation function vanishes in the Gaussian limit where $\mu \to \infty$, with $a \to 0$.

- **Exercise:** Consider an amplitude- and phase-modulated narrow-frequency-band signal

$$s(t) = R(t)\cos(\omega_c t + \theta(t)).$$

An equivalent representation is

$$s(t) = x(t)\cos(\omega_c t) - y(t)\sin(\omega_c t).$$

Suppose that the joint probability density $P(x, y)$ is known. Derive an expression for the joint probability density function $P(R, \theta)$, and write the expression for the marginal distributions $P(R)$ and $P(\theta)$.

Assume now that $s(t)$ is a normally distributed random signal

$$s(t) = \sum_{n=1}^{\infty} a_n \cos(n\omega t) + \sum_{n=1}^{\infty} b_n \sin(n\omega t),$$

where $\langle a_n b_m \rangle$ for all n and m, while $\langle a_n a_m \rangle = \langle b_n b_m \rangle = 0$ for $n \neq m$, and $\langle a_n^2 \rangle = \langle b_n^2 \rangle = \sigma^2$. Demonstrate that then also $x(t)$ and $y(t)$ are normally distributed. Given the foregoing results, demonstrate that $P(R)$ has a Rayleigh distribution

$$P(R) = \frac{R}{\sigma^2} e^{-R^2/(2\sigma^2)} \tag{9.39}$$

for $R > 0$ and zero otherwise. Note that (9.39) is independent of ω_c. Determine $\langle R \rangle$ and $\langle R^2 \rangle$. What is the distribution of θ? The Rayleigh distribution is shown in Fig. 9.3.

9.5   Applications

The present model (9.1) can give an excellent representation for a variety of phenomena in systems that are not necessarily in thermal equilibrium. As an example we could model the noise caused by raindrops falling on a tin roof. A more common, but prosaic, application is for shot noise in diodes. In this case it can be argued that a model basic triangular current pulse is adequate as an illustration

$$i(t) = 2et/\tau_\alpha^2$$

for $0 < t < \tau_\alpha$ and $i(t) = 0$ otherwise (Davenport and Root, 1958). The auto-correlation function is readily obtained as

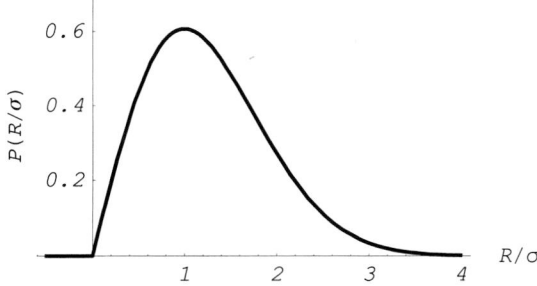

Figure 9.3 The probability density for the normalized Rayleigh distribution $P(R/\sigma)$, as given by (9.39).

$$R_i(\tau) = \frac{4e\langle I\rangle}{3\tau_\alpha}\left(1 - \frac{3}{2}\frac{|\tau|}{\tau_\alpha} + \frac{1}{2}\frac{|\tau|^3}{\tau_\alpha^3}\right) + \langle I\rangle^2$$

for $|\tau| < \tau_\alpha$ and $R_i(\tau) = \langle I\rangle^2$ otherwise, with $\langle I\rangle = e\mu$, where here μ is the number of electrons arriving at the cathode per second. By Fourier transforming the correlation function we obtain the frequency spectrum

$$S_i(\omega) = e\langle I\rangle \frac{4}{(\omega\tau_\alpha)^4}\{(\omega\tau_\alpha)^2 + 2[1 - \cos(\omega\tau_\alpha) - \omega\tau_\alpha\sin(\omega\tau_\alpha)]\},$$

apart from a delta-function at the origin due to the direct current. The normalized correlation function and corresponding power spectrum are shown in Fig. 9.4. In particular, the low-frequency limit of the spectral density of the shot-noise current generated by a temperature-limited diode is $S_i(\omega) = e\langle I\rangle = e^2\mu$, known as the *Schottky formula*. Modifications to these results due to space-charge limitations are discussed in the literature (Davenport and Root, 1958, MacDonald, 1962). In particular, a space-charge-smoothing factor Γ^2 with $0 < \Gamma^2 \le 1$ is introduced into the Schottky formula. Note that the example in this section concerns a system *out of* thermal equilibrium and the time stationarity of the process is maintained by external sources ensuring the diode's operation.

More generally, we can describe a different physical situation in which two, or more, signal sources are present, each of them randomly emitting wave packets $\psi(t)$, the output being adequately represented by the model (9.1). Assume for simplicity that the signal generators have the same statistical properties. Evidently, the resulting signal obtained by superposition of the outputs from the various sources will then have the same basic statistical properties as those

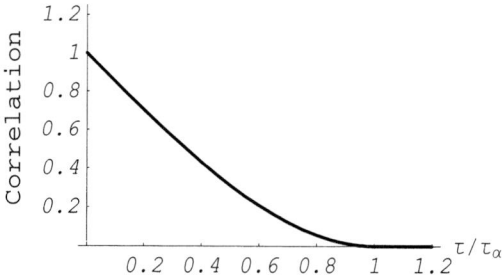

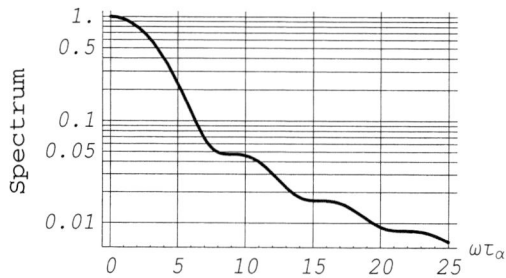

Figure 9.4 The correlation function, $(R_i(\tau) - \langle I\rangle^2)\tau_\alpha/e\langle I\rangle$, and power spectrum, $S_i(\omega\tau_\alpha)/e\langle I\rangle$, obtained with a random superposition of triangular pulses. Note that there are local maxima in the power spectrum, but no oscillations in the correlation function.

of a single source, with the exception that the density μ has to be interpreted as $M\mu_s$, where M is the number of sources and μ_s is the density of pulses, or wave packets, associated with one source, labeled by s. A basic consequence of this almost trivial observation is that, for instance, the power associated with the signal is proportional to the number of signal sources; five violins are five times as loud as one! (Sometimes it is experienced as being worse than that, but this is a subjective opinion without basis in the analysis.) This is actually not quite a trivial observation, for it is the amplitudes which are added by superposition of the signals, while the power is obtained by squaring the entire signal.

The light from a group of statistically independent incoherent light sources can be discussed by deploying entirely similar arguments. If the light sources have different spectra, the spectrum of their superimposed signal is readily obtained from the analysis as the weighted sum of the individual spectra.

- **Example:** For incoherent radar scattering from the ionosphere (Bekefi, 1966), we can to a first approximation assume that all electrons scatter the incoming radar beam independently. Some contributions cancel out while others add up, just like for a random superposition of wave packets in a sequence such as (9.1). The net scattering intensity will be proportional to the number of scatterers (i.e. electrons) in the scattering volume.

9.6   White noise

In particular a signal with a constant power spectrum, a *white-noise* signal, can be realized by the model (9.1), by choosing the basic pulse to be a δ-function. In that case we readily find the correlation function by the convolution

$$R(\tau) = \mu a^2 \int \delta(t)\delta(t + \tau)\,dt = \mu a^2 \delta(\tau),$$

with a an amplitude and μ the pulse density, as before. Fourier transformation gives the constant power spectrum $G(\omega) = 2\mu a^2/\pi$. The total energy in this spectrum is infinite, because the correlation function is infinite at the origin, $\tau = 0$, implying that the mean-square value of the signal diverges.

9.7   1/*f* noise

It may be noted that most of the models discussed in this chapter and in particular also in Chapter 3 have a 'white' (or constant) frequency spectrum as $\omega \to 0$ (although it *is* possible to construct simple models in which $S(\omega) \to 0$ for $\omega \to 0$). In actual experiments it is, however, often observed that the spectral power increases approximately as $1/\omega$ (i.e. $1/f$) for small ω. For instance Johnson (1925) studied the fluctuations in current for electronic emission in thermionic tubes, and found that, apart from the shot noise, for which the spectral power density was constant, independent of frequency, in agreement with Schottky's formula, there was also a component with increasing power for decreasing frequency. Schottky (1926) suggested that this noise arose from random changes in the surface of the thermo cathode, and called this a 'flicker effect' of 'flicker noise.' It eventually became evident that this type of noise is a phenomenon that is very often encountered, if not universal, in conductors (Kogan, 1996).

As a basic model that might explain the occurrence of $1/f$ noise, we here consider a fluctuating current $i(t)$, with a correlation function $\langle i^2 \rangle \exp(-|\tau|/\tau_c)$, corresponding to emission of electrons from a hot surface as in Section 9.5. The power spectrum is then the Lorentzian function of frequency

$$S(\omega) = \langle i^2 \rangle \frac{4\tau_c}{1 + (\omega\tau_c)^2}. \tag{9.40}$$

The constants $\langle i^2 \rangle$ and τ_c in this expression refer to one particular surface condition of the cathode. Assume now that the surface is composed of many surface elements with slightly varying conditions; after all, this is the most realistic case. The work function of a material typically depends on the crystal surface facing the free space, just to mention one example, and a realistic cathode is unlikely to be one large single crystal, so the surface contains contributions from many different work functions, and surface conditions. Assume for instance that the mean-square current contribution contains many independent simultaneous contributions with different relaxation times τ_c, with a known statistical distribution $P(\tau_c)$. Averaging over all τ_c, we have trivially

$$S(\omega) = \langle i^2 \rangle \int_0^\infty \frac{4\tau_c}{1 + (\omega\tau_c)^2} P(\tau_c) d\tau_c. \tag{9.41}$$

If now $P(\tau_c) \sim 1/\tau_c$ in an interval $\{\tau_1; \tau_2\}$ and vanishes otherwise, we find that

$$S(\omega) = \frac{\langle i^2 \rangle}{\omega} \frac{4}{\ln(\tau_2/\tau_1)} \int_{\omega\tau_1}^{\omega\tau_2} \frac{d\gamma}{1 + \gamma^2}, \tag{9.42}$$

implying that $S(\omega) \sim 1/\omega$ in the frequency interval $\tau_1^{-1} \ll \omega \ll \tau_2^{-1}$.

By proper choice of $P(\tau_c)$ it is thus possible to account for the $1/\omega$ noise in this case, but of course the physical origin of precisely this distribution for $P(\tau_c)$ remains to be determined! However, this discussion is problematic also for a quite different reason; $1/\omega$ noise seems to occur quite generally, and the foregoing arguments are too specific. $1/\omega$ noise is encountered for instance in currents through carbon resistors (Moore, 1974) and metal films (Clarke and Voss, 1974), but also in flows of particles through narrow pores (Shick and Verveen, 1974). Clearly, a *universal* theory is to be sought. One basic, heuristic, almost 'hand-waving' argument has been suggested: white noise can be explained by invoking a random superposition of many narrow pulses, in effect δ-pulses. If the coefficient of the power spectrum were to vary as 'step functions' with random, but large, intervals, one might expect the low-frequency limit to behave as the Fourier transform of a step function, i.e. $1/\omega$. (Note that we are not arguing that there is a random distribution of individual pulses such as in a random telegraph wave, for instance. This will give $1/\omega^2$ for low frequencies, although not in the limit of $\omega \to 0$!) Recent developments in studies of this and related phenomena introduced the concept of 'self-organized criticality' (Bak *et al.*, 1987, Bak, 1997). In agreement with many observations, it is claimed that slowly driven and strongly interacting systems undergo self-organization, without any external influence, until they reach an extremely sensitive state, whereupon very small, local, disturbances can give rise to avalanches of intermittent events on all scales. Spectral analysis of such systems reveals that they follow an $f^{-\beta}$ power law, with β very close to 1. In many ways, the behavior of such systems may visually resemble long queues of cars on an express highway with randomly distributed stop lights. Indeed, a $1/f$ power law has been observed for the fluctuations in traffic flow on highways (Musha and Higuchi, 1976).

The lower frequency limit for $1/f$ noise is not well accounted for. Actually, one puzzle in this field is the apparent divergence of the integrated power when $\omega \to 0$.

The discussions in this subsection are meant as an outline only. A detailed discussion of $1/f$ noise is beyond the scope of this book.

9.8   Consequences of finite record lengths

The actual estimates of the correlations for a model such as (9.1) from a given record are subject to uncertainties due to the finite lengths of records. The model allows expressions for these uncertainties to be obtained also (Bendat, 1958, Pécseli and Trulsen, 1993).

9.8.1   Averages

For instance, assume that in the model we consider a case in which the average value for the signal vanishes, $\langle \Phi(t) \rangle = \mu a \int_{-\infty}^{\infty} \psi(t)\, dt = 0$. An actual estimate of this average, $(1/T) \int_0^T \psi(t)\, dt$, based on a finite sample length, T, will in general be different than zero. The estimate can itself be considered as a random variable and its standard deviation σ_{av} obtained as

$$\sigma_{av}^2 = \left\langle \left(\frac{1}{T} \int_0^T \Phi(t)\, dt - \langle \Phi(t) \rangle \right)^2 \right\rangle. \tag{9.43}$$

With $\int_{-\infty}^{\infty} \psi(t)\, dt = 0$ this expression becomes

$$\sigma_{av}^2 = \frac{2}{T^2} \int_0^T (T - \tau) \mathcal{R}(\tau)\, d\tau, \tag{9.44}$$

indicating that $\sigma_{av} \to 0$ as $T \to \infty$, i.e. when T is much larger than the correlation time. On the other hand, for small T we have $\sigma_{av}^2 \approx R(0) = \langle \Phi^2 \rangle$.

9.8.2   Auto-correlation functions

The estimate $(1/T) \int_0^T \Phi(t)\Phi(t + \tau)\, dt$ for the auto-correlation function is also randomly varying over the ensemble of many finite length records. The standard deviation for the error of this estimate is

$$\sigma_{\text{corr}}^2(\tau) = \left\langle \left(\frac{1}{T} \int_0^T \Phi(t)\Phi(t+\tau)\, dt - \langle \Phi(t)\Phi(t+\tau)\rangle \right)^2 \right\rangle$$

$$= \frac{2}{T^2} \int_0^T (T-v)\big[\mathcal{Q}(\tau, v+\tau, v) - \mathcal{R}^2(\tau)\big]\, dv, \qquad (9.45)$$

where $\mathcal{Q}(\tau_1, \tau_2, \tau_3) = \langle \Phi(t)\Phi(t+\tau_1)\Phi(t+\tau_2)\Phi(t+\tau_3)\rangle$ is the quadruple-correlation function. For the model given by (9.1) we obtain analyticly

$$\mathcal{Q}(\tau_1, \tau_2, \tau_3) = \mu a^4 \int_{-\infty}^{\infty} \psi(t)\psi(t+\tau_1)\psi(t+\tau_2)\psi(t+\tau_3)\, dt$$

$$+ \mathcal{R}(\tau_1)\mathcal{R}(\tau_3 - \tau_2) + \mathcal{R}(\tau_2)\mathcal{R}(\tau_3 - \tau_1) + \mathcal{R}(\tau_3)\mathcal{R}(\tau_1 - \tau_2). \qquad (9.46)$$

We again explicitly used that $\langle \Phi(t)\rangle = 0$. The term $\mathcal{R}^2(\tau)$ cancels out when (9.45) is inserted into (9.46). We note that $\sigma_{\text{corr}}^2(\tau)$ contains $(\mu a^2)^2$ terms, originating from the second term in (9.46), and one varying with $(\mu a^2)a^2$, the latter vanishing in the Gaussian limit. The error in the mean-square estimate can be obtained as the special case with $\tau = 0$ in (9.45).

- **Exercise:** Demonstrate that

$$\mathcal{Q}(\tau, v+\tau, v) = \mathcal{R}^2(\tau) + \mathcal{R}^2(v) + \mathcal{R}(\tau+v)\mathcal{R}(\tau-v)$$

  for a Gaussian random process described by its correlation function $\mathcal{R}(\tau)$.

- **Example:** Considering a Gaussian random process with the correlation function $R(\tau) = e^{-b\tau}$, we can simplify the expression for $\mathcal{Q}$ in (9.46). It follows that

$$\mathcal{Q}(\tau, v+\tau, v) - \mathcal{R}^2(\tau) = e^{-2b|v|} + e^{-b|\tau+v|}e^{-b|\tau-v|}.$$

  Then (9.45) breaks up into two integrals:

$$\sigma_{\text{corr}}^2(\tau) = \frac{2}{T^2} \int_0^\tau (T-v)\left(e^{-bv} + e^{-b\tau}\right) dv + \frac{4}{T^2} \int_0^\tau (T-v)e^{-2bv}\, dv.$$

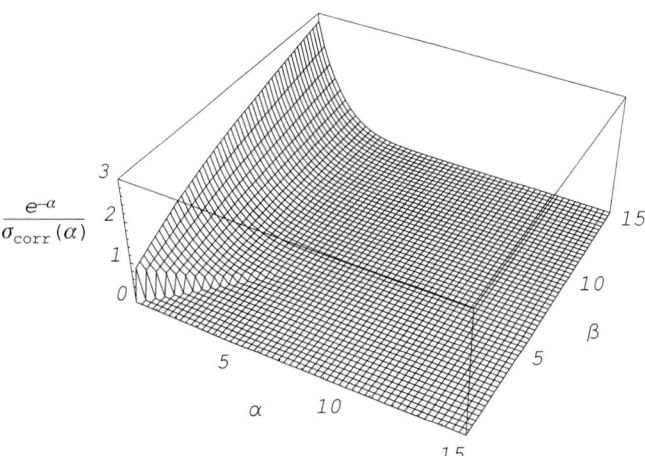

Figure 9.5 The signal-to-noise ratio for an estimate of a particular correlation function, here given as the ratio $e^{-\alpha}/\sigma_{\text{corr}}(\alpha)$, see (9.47), shown as a function of normalized variables $\alpha = b\tau$ and $\beta = bT$.

On introducing dimensionless variables $\alpha = b\tau$ and $\beta = bT$, we find that

$$\sigma^2_{\text{corr}}(\alpha) = \frac{1}{2\beta^2}\{2\beta - 1 + 2e^{-2\beta} + [(2\alpha + 1)(2\beta - 1) - 2\alpha^2]e^{-2\alpha}\}. \quad (9.47)$$

This relation is exact for all α and $\beta > \alpha \geq 0$, that is for $T > \tau \geq 0$. For a fixed value of β, the variance σ_{corr} indicates an uncertainty in the knowledge of $R(\tau)$ for all τ. For $\alpha = 0$, the result (9.47) reduces to

$$\sigma^2_{\text{corr}}(0) = \frac{2\beta - 1 + e^{-2\beta}}{\beta^2}, \quad (9.48)$$

giving $\sigma^2_{\text{corr}}(0) \approx 2/\beta$ for large β. The 'signal-to-noise ratio' for this problem can generally be expressed as $R(\tau)/\sigma_{\text{corr}}(\tau)$, which in the present case is $\exp(-\alpha)/\sigma_{\text{corr}}(\alpha)$. This function is shown in Fig. 9.5. For large β we find, using (9.47), that $e^{-\alpha}/\sigma_{\text{corr}}(\alpha) \approx e^{-\alpha}\sqrt{\beta/k}$, with k a number in the range $\{1; 2\}$ (Bendat, 1958). For $\alpha = 0$, in particular, we find a $\sqrt{\beta}$ dependence, as expected. The signal-to-noise ratio as defined here should be large for reliable estimates, and Fig. 9.5 shows that this requires long, perhaps surprisingly long, time series, in particular if $\tau > 0$. Although this example refers to particular conditions, it is a useful illustration also for more general cases.

9.8.3   Triple-correlation functions

Also the estimate for a triple-correlation function derived from a time series of finite length will deviate from the exact analytic result. An expression of the standard deviation of the error in such an estimate of the triple-correlation function can be obtained as

$$\sigma^2_{\text{trp}}(\tau_1, \tau_2) = \left\langle \left(\frac{1}{T}\int_0^T \Phi(t)\Phi(t + \tau_1)\Phi(t + \tau_2)\,dt - \langle \Phi(t)\Phi(t + \tau_1)\Phi(t + \tau_2)\rangle\right)^2 \right\rangle$$

$$= \frac{2}{T^2}\int_0^T (T - v)[\mathcal{X}(\tau_1, \tau_2, v, v + \tau_1, v + \tau_2) - \mathcal{S}^2(\tau_1, \tau_2)]\,dv, \quad (9.49)$$

expressed in terms of the sixth-order correlation function

$$\mathcal{X}(\tau_1, \tau_2, \tau_3, \tau_4, \tau_5) = \mu a^6 \int_{-\infty}^{\infty} \psi(t)\psi(t + \tau_1)\psi(t + \tau_2)\psi(t + \tau_3)\psi(t + \tau_4)\psi(t + \tau_5)\,dt$$

$$+ \sum_{\text{perm}(h,i,j,k,l)}^{15} \mathcal{R}(\tau_h)\mathcal{R}(\tau_i - \tau_j)\mathcal{R}(\tau_k - \tau_l)$$

$$+ \sum_{\text{perm}(h,i,j,k,l)}^{15} \mathcal{R}(\tau_h)\mathcal{Q}(\tau_i - \tau_j, \tau_k - \tau_j, \tau_l - \tau_j)$$

$$+ \sum_{\text{perm}(h,i,j,k,l)}^{15} \mathcal{S}(\tau_h, \tau_i)\mathcal{S}(\tau_j - \tau_l, \tau_k - \tau_l). \quad (9.50)$$

The summations are to be extended over all permutations of the τ indices, without double counting, just like with the last term in (9.46). Insertion of (9.50) into (9.49) results in cancellation of the $\mathcal{S}^2(\tau_1, \tau_2)$ term. A term proportional to $(\mu a^2)^3$ appears in the error estimate $\sigma_{\text{trp}}^2(\tau_1, \tau_2)$. This term, which originates from the second term in (9.50), will remain in the Gaussian limit. In addition we find error contributions varying as $(\mu a^2)^2 a^2$ originating from the two last terms in (9.50).

9.8.4   Cross-correlation functions

The consequences of finite record lengths have been analyzed for cross-correlations as well. The result is particularly interesting when two statistically independent records are analyzed. Ideally, the correlation should vanish, but in practice it will assume some nonzero value due to the finite length of the data sequence. Actually, it is rather difficult to find two records that give a perfectly vanishing cross-correlation, e.g. the birth rate of babies and the population of storks in Denmark are found to be correlated.

The moral of the analysis in this section is that the uncertainty in the estimation for a certain statistical moment, i.e. a correlation function of some order, depends on higher order correlations than the one investigated. It is not logically possible to determine these uncertainties experimentally; to express σ_{av}^2 we need $R(\tau)$, which can be measured only with some uncertainty, σ_{corr}^2, which can only be obtained with knowledge of $\mathcal{Q}(\tau_1, \tau_2, \tau_3)$, which is again uncertain, etc. From an experimental point of view, the uncertainties σ_{av}^2, σ_{corr}^2, etc. are themselves also statistically uncertain quantities! Analytically, they can be obtained only by assuming certain models for the probabilities. These models build on some additional knowledge (or assumed knowledge) concerning the processes which may be supported, but never proven, by measurements carried out on the system.

10 Markov processes

Random processes of a particular type, Markov processes, play an important role in natural phenomena. At least it can be argued that they give an adequate approximation for a statistical analysis of many natural phenomena. In words, these processes can be described in terms of a lack of memory; knowledge of the actual state of systems described by these processes determines the distribution of future states (Papoulis, 1991). For Markovian systems the present separates the past from the future; knowledge of the system at one instant, the present, permits prediction of the 'average' future and indeed the probabilities of various possible futures (Lax, 1960, Cox and Miller, 1977). Information about the past is not needed, and is not relevant even when it is available. A simple example of a Markov process was presented in Section 3.2. Here the basic properties of such systems will be discussed without particular reference to actual physical details. In particular it is demonstrated that the relaxation to thermal equilibrium of Markovian systems in response to an initial perturbation can be given a general formalism.

It is advantageous to discuss Markov processes in terms of probability densities (Cox and Miller, 1977). Let $Y_x(t)$ be a stochastic process, i.e. an ensemble of functions of t labeled by the parameter x with a given distribution $D(x)$. As before, we use the notation Y for the variable and lower-case y for the values it can assume.

Let $P_1(y_1, t_1) \, dy_1$ be the probability that $Y(t_1)$ lies in the interval $[y_1; y_1 + dy_1]$. Then

$$P_1(y_1, t_1) = \int_{-\infty}^{\infty} \delta[Y_x(t_1) - y_1] D(x) \, dx. \tag{10.1}$$

Similarly, let $P_2(y_1, t_1; y_2, t_2) \, dy_1 \, dy_2$ be the joint probability that $Y(t_1)$ lies in the interval $[y_1; y_1 + dy_1]$ and simultaneously $Y(t_2)$ lies in the interval $[y_2; y_2 + dy_2]$. In this way one can define a hierarchy of probabilities $P_1, P_2, P_3, \ldots$ that constitutes a possible way of describing the stochastic process. One can also define conditional probability densities in terms of these joint probability density functions

$$P_{n|n-1}(y_n, t_n | y_1, t_1; \ldots, y_{n-1}, t_{n-1}) = \frac{P_n(y_1, t_1; \ldots; y_n, t_n)}{P_{n-1}(y_1, t_1; \ldots; y_{n-1}, t_{n-1})} \tag{10.2}$$

by use of Bayes' rule, where the time ordering $t_n \geq t_{n-1} \geq \cdots \geq t_2 \geq t_1$ was assumed. In the following we omit the index on conditional probabilities for simplicity.

The Markov property is expressed in terms of the conditional probabilities as

$$P(y_n, t_n | y_1, t_1; \ldots; y_{n-1}, t_{n-1}) = P(y_n, t_n | y_{n-1}, t_{n-1}), \tag{10.3}$$

with a self-evident interpretation, i.e. the state y_{n-1}, t_{n-1} determines the probability density of states y_n, t_n irrespective of the states y_m, t_m with $m < n - 1$. An arbitrary joint probability density for a Markov process can then be expressed as

$$P_n(y_1, t_1; \ldots, y_n, t_n) = P(y_n, t_n | y_{n-1}, t_{n-1}) \cdots P(y_2, t_2 | y_1, t_1) P_1(y_1, t_1). \tag{10.4}$$

For Markov processes all information is contained in the joint probability density function, P_2, since P_1 can be derived from it together with the conditional probability density P_2/P_1 by use of Bayes' rule, (2.14). In particular we also have

$$P(y_2, t_2; y_3, t_3 | y_1, t_1) = P(y_3, t_3 | y_2, t_2; y_1, t_1) P(y_2, t_2 | y_1, t_1). \tag{10.5}$$

With the Markov property this relation becomes

$$P(y_2, t_2; y_3, t_3 | y_1, t_1) = P(y_3, t_3 | y_2, t_2) P(y_2, t_2 | y_1, t_1). \tag{10.6}$$

In terms of conditional probabilities we have

$$\begin{aligned} P_1(y_2, t_2) &= \int_{-\infty}^{\infty} P_2(y_1, t_1; y_2, t_2) \, dy_1 \\ &= \int_{-\infty}^{\infty} P(y_2, t_2 | y_1, t_1) P_1(y_1, t_1) \, dy_1, \end{aligned} \tag{10.7}$$

and also

$$\begin{aligned} P(y_3, t_3 | y_1, t_1) &= \int_{-\infty}^{\infty} P(y_2, t_2; y_3, t_3 | y_1, t_1) \, dy_2 \\ &= \int_{-\infty}^{\infty} P(y_3, t_3 | y_2, t_2; y_1, t_1) P(y_2, t_2 | y_1, t_1) \, dy_2. \end{aligned} \tag{10.8}$$

On introducing the Markov assumption this relation becomes

$$P(y_3, t_3 | y_1, t_1) = \int_{-\infty}^{\infty} P(y_3, t_3 | y_2, t_2) P(y_2, t_2 | y_1, t_1) \, dy_2, \tag{10.9}$$

which is known as the Smoluchowski equation or the Chapman–Kolmogorov equation. It states that the probability of going from y_1 to y_3 via y_2 is the product of the probability for going from y_1 to y_2 times that for going subsequently from y_2 to y_3; that is, successive transitions are statistically independent. The time ordering $t_1 < t_2 < t_3$ is essential. It can *not* be assumed that an arbitrary joint probability density can describe a Markov process; it has to satisfy (10.9) also.

10.1   The master equation

A more manageable form of (10.9) is obtained by taking $t_3 = t_2 + \Delta t$ with small Δt. For physically acceptable processes it is expected that the probability that Y makes a jump during Δt will be proportional to Δt apart from small corrections. This is expressed by

$$P(y_3, t_3 | y_2, t_2) = \delta(y_3 - y_2)[1 - w(y_2) \Delta t] + \Delta t \, W(y_3 | y_2) + \mathcal{O}(\Delta t^2), \tag{10.10}$$

where $W(y_3 | y_2)$ is the probability per unit time of transition from y_2 to y_3. The term $w(y_2) \Delta t$ is the total probability that a transition *from* y_2 to *any* y_3 took place, i.e. $w(y_2) = \int W(y_3 | y_2) \, dy_3$, while $1 - w(y_2) \Delta t$ is the probability of nothing happening in the time interval Δt. Ignoring terms $\mathcal{O}(\Delta t^2)$ in (10.10), direct substitution gives

$$\frac{\partial P(y, t | y_1, t_1)}{\partial t} = \int_{-\infty}^{\infty} [W(y | y') P(y', t | y_1, t_1)$$
$$- W(y' | y) P(y, t | y_1, t_1)] \, dy', \tag{10.11}$$

with use of the approximation

$$\frac{\partial P(y, t | y_1, t_1)}{\partial t} \approx \frac{P(y, t | y_1, t_1) - P(y, t - \Delta t | y_1, t_1)}{\Delta t}.$$

As an alternative, more customary formulation, select a sub-ensemble of the functions $Y_x(t)$ by prescribing a distribution $P(y_1, t_1)$ at some initial time t_1. Then the distribution in this sub-ensemble at $t \geq t_1$ obeys the *master equation*

$$\frac{\partial P(y, t)}{\partial t} = \int_{-\infty}^{\infty} [W(y | y') P(y', t) - W(y' | y) P(y, t)] \, dy'. \tag{10.12}$$

This differential version of the *Chapman–Kolmogorov equation*, which is valid for the probability density associated with any stationary Markov process obeying (10.10), is called the master equation. For a discrete sample space but still a continuous time variable, this equation takes the form

$$\frac{\partial P_n(t)}{\partial t} = \sum_{n'} [W_{n,n'} P_{n'}(t) - W_{n',n} P_n(t)]. \tag{10.13}$$

These equations have an intuitively plausible appearance; they are gain–loss equations for the probability of states y or, in the discrete case, n. The first term is the gain due to transitions from other states y', and the second term is the loss due to transitions into other states. It is advantageous to make further simplifications of the master equations. The Fokker–Planck equation is a particularly useful example.

10.2   Fokker–Planck equations

As a first step the transition probability is rewritten in terms of the size r of the jump and of the starting point y as $W(y | y') = W(y'; r)$ with $r = y - y'$. The master equation can then be written as (van Kampen, 1981)

$$\frac{\partial P(y, t)}{\partial t} = \int_{-\infty}^{\infty} W(y - r; r) P(y - r, t) \, dr - P(y, t) \int_{-\infty}^{\infty} W(y; -r) \, dr. \tag{10.14}$$

The basic assumption is now that only small jumps occur, that is $W(y'; r)$ is a sharply peaked function of r but varies slowly with y. The second assumption is that $P(y, t)$ varies only slowly with y for cases of interest. By introducing a Taylor expansion up to second order, it is possible to shift from $y - r$ to y in the first integrand:

$$\frac{\partial P(y, t)}{\partial t} = \int_{-\infty}^{\infty} W(y; r) P(y, t) \, dr - \int_{-\infty}^{\infty} r \frac{\partial}{\partial y} [W(y; r) P(y, t)] \, dr$$
$$+ \frac{1}{2} \int_{-\infty}^{\infty} r^2 \frac{\partial^2}{\partial y^2} [W(y; r) P(y, t)] \, dr$$
$$- P(y, t) \int_{-\infty}^{\infty} W(y; -r) \, dr. \tag{10.15}$$

The first and last terms cancel out, as is demonstrated by a change of integration variable, while the remaining two terms are expressed by introducing the jump moments

$$a_n(y) = \int_{-\infty}^{\infty} r^n W(y; r)\, dr, \tag{10.16}$$

to give the Fokker–Planck equation

$$\frac{\partial P(y, t)}{\partial t} = -\frac{\partial}{\partial y}[a_1(y)P(y, t)] + \frac{1}{2}\frac{\partial^2}{\partial y^2}[a_2(y)P(y, t)], \tag{10.17}$$

wherein only a_1 and a_2 enter. For an n-dimensional process equation (10.17) is generalized to

$$\frac{\partial P}{\partial t} = -\sum_l \frac{\partial}{\partial y_l}[A_l(y)P] + \frac{1}{2}\sum_{l,m}\frac{\partial^2}{\partial y_l \partial y_m}[B_{lm}(y)P], \tag{10.18}$$

where now A_l and B_{lm} are the components of a vector and a tensor, respectively (van Kampen, 1981).

The Fokker–Planck equation allows a study of nonstationary conditions, i.e. the temporal evolution of the probability density function, in particular the relaxation to thermal equilibrium. The diffusion equation is a particularly simple version. Often the Fokker–Planck equation is solved with the initial condition $P(y, 0) = \delta(y)$, although strictly speaking this function violates the assumption of P being slowly varying with y.

In principle one might include all terms in a Taylor expansion of (10.14), giving

$$\frac{\partial P(y, t)}{\partial t} = \sum_{j=1}^{\infty}\frac{(-1)^j}{j!}\frac{\partial^j}{\partial y^j}[a_j(y)P(y, t)],$$

which is known as the Kramers–Moyal expansion, of which a detailed discussion has been given for instance by van Kampen (1981). The Fokker–Planck equation assumes that all terms with $j > 2$ are negligible in the Kramers–Moyal expansion. Note that this assumption need not always be justified (van Kampen, 1981).

- **Exercise:** Demonstrate that the form

$$\frac{\partial P}{\partial t} = -\sum_l C_l \frac{\partial}{\partial y_l}P - \sum_l A_l \frac{\partial}{\partial y_l}(y_l P) + \frac{1}{2}\sum_{l,m}B_{lm}\frac{\partial^2}{\partial y_l \partial y_m}P \tag{10.19}$$

    can be reduced to (10.18) by an appropriate change of variables (van Kampen, 1981).

- **Example:** The summary in Sections 10.1 and 10.2 was formulated in configuration space, but evidently there is no particular reason for restricting the analysis in this way; since we can imagine the heating of a gas of particles as a random walk in velocity space, we might write a Fokker–Planck equation in velocity space, or more generally in phase space. In (10.17) we can therefore replace the position variable y by the velocity u and obtain the new coefficients a_1 and a_2 by use of the relevant transition probabilities. The meaning of the time differential operator should now be examined carefully, however. Generally it should be interpreted as the usual Stokes operator $D/Dt \equiv \partial/\partial t + \boldsymbol{u} \cdot \nabla + \boldsymbol{K} \cdot \nabla_u$, where $\boldsymbol{K}$ is an acceleration due to some (here unspecified) external forces, and $\nabla_u \equiv (\partial/\partial u_x)\hat{x} + (\partial/\partial u_y)\hat{y} + (\partial/\partial u_z)\hat{z}$ is the velocity derivative operator (Chandrasekhar, 1943). We have

$$\frac{\partial}{\partial t}P + u \cdot \nabla P + K \cdot \nabla_u P = A\nabla_u \cdot (uP) + B\nabla_u^2 P, \tag{10.20}$$

which is the generalization of the Fokker–Planck equation and also of Liouville's theorem of classical mechanics. Here, we assumed A and B to be constants for simplicity.

Physically, (10.20) accounts for the increments of a particle's position and velocity $\Delta r = u\,\Delta t$ and $\Delta u = -(Au - K/M)\,\Delta t + (1/M)F(\Delta t)$, respectively, in a small time interval Δt, with A giving a 'friction' in phase space, and K/M an acceleration of particles with mass M, while $F(\Delta t)$ is the fluctuating force the particle experiences by virtue of its interaction with the surroundings, the distribution of F being assumed to be a Gaussian, determining B in (10.20). In general, K can depend on space and time. In thermal equilibrium, a spatially homogeneous steady-state solution of (10.20) with $K = 0$ satisfies the vector relation $AuP + B\nabla_u P = 0$, giving $P = [A/(2\pi B)]^{3/2}\exp[-\frac{1}{2}(A/B)u^2]$. From basic thermodynamics we require this solution to be a Maxwellian, $[M/(2\pi\kappa T)]^{3/2}\exp[-Mu^2/(2\kappa T)]$, imposing the condition $A/B = M/(\kappa T)$. Considering for instance high-frequency electrostatic waves in a plasma in thermal equilibrium, the term $B\nabla_u^2 P$ in (10.20) can be interpreted as a heating, giving rise to a diffusion or broadening of the distribution function in velocity space, such that the heating is brought about by the acceleration of the electric field of the thermally excited Langmuir waves (4.25). The term $A\nabla_u \cdot (uP)$ can then be interpreted as the friction or loss of energy due to radiation (similar to Čerenkov radiation) of Langmuir waves by fast charged particles in the plasma. The Maxwell distribution arises when the two phenomena balance. Note that the form

$$\frac{\partial P}{\partial t} = -u\frac{\partial}{\partial x}P + x\frac{\partial}{\partial u}P - \gamma\left(\frac{\partial}{\partial u}(uP) + \frac{\partial^2}{\partial^2 u}P\right)$$

(a bivariate Fokker–Planck equation) is contained in (10.19).

10.2.1   Transition from a discrete to a continuous case

A discrete random walk of a free particle can in a limiting case be modeled by using a Fokker–Planck equation. It is instructive to see how this transition from a discrete to a continuous case is brought about. Imagine a particle that moves along the x-axis in such a way that in each step it can move either Δx to the left or Δx to the right. For a free particle we can assume that a step in either of the two directions is equally probable, with probability $\frac{1}{2}$. Consider $P(n\,\Delta x\,|\,m\,\Delta x;\,s\tau)$, which is the probability density that the particle is at $m\,\Delta x$ at time $s\tau$, if at the beginning it was at $n\,\Delta x$. The conditional probability density satisfies the difference equation (Kac, 1946)

$$P(n|m; s+1) = \tfrac{1}{2}P(n|m+1; s) + \tfrac{1}{2}P(n|m-1; s), \tag{10.21}$$

which can be rewritten in an equivalent form as

$$\frac{P(n\,\Delta x|m\,\Delta x; (s+1)\tau) - P(n\,\Delta x|m\,\Delta x; s\tau)}{\tau} =$$

$$\frac{\Delta x^2}{2\tau}\left(\frac{P(n\,\Delta x|(m+1)\,\Delta x; s\tau) + P(n\,\Delta x|(m-1)\,\Delta x; s\tau)}{\Delta x^2}\right.$$

$$\left. - \frac{2P(n\,\Delta x|m\,\Delta x; s\tau)}{\Delta x^2}\right). \tag{10.22}$$

Suppose now that $\Delta x \to 0$ and $\tau \to 0$ in (10.22) in such a way that $\Delta x^2/2\tau = D = $ constant. In this limit, with $n\,\Delta x \to x$ and $s\tau \to t$, we recover the diffusion equation from (10.22) as

$$\frac{\partial P}{\partial t} = D\frac{\partial^2 P}{\partial x^2},$$

where D is a diffusion coefficient.

An important implication of the transition allowed to obtain this equation is that the instantaneous particle velocity $\Delta x/\tau$ approaches infinity. This explains why the diffusion equation has a finite response at any position, no matter how far from the origin, at arbitrarily short times, and also for initial conditions with finite support. With the local particle velocity being infinite, we also find that the length of the path (not the net displacement) of a diffusing particle diverges in this diffusion limit.

The arguments outlined in this section contain two elements; first time is made continuous by using infinitesimal timesteps and, simultaneously, the spatial variable is also made continuous by using very small steps or jumps. This limit can be obtained by 'brute force' using (10.16) with the transition probability $W_{n,n'} = (1/\tau)(\frac{1}{2}\Delta_{n',n+1} + \frac{1}{2}\Delta_{n',n-1})$ per time interval τ for the symmetric random walk. Clearly, $a_1 = 0$ while $a_2 = \frac{1}{2}\Delta x^2/\tau \equiv D$ and the corresponding Fokker–Planck equation becomes again the diffusion equation, $\partial P/\partial t = D\,\partial^2 P/\partial x^2$. With some care, the expression (10.16) can thus be used also to obtain a Fokker–Planck equation for processes that are by nature discrete, with some self-evident restrictions imposed on the interpretation of the results.

The limiting process of a simple random walk is called the *Wiener process*, or the *Brownian-motion process*. For such a process, the increment $\Delta X(t) = X(t+\Delta t) - X(t)$ in a small time interval Δt is independent of $X(t)$ and has a mean and a variance proportional to Δt. The process is a Markov process, since, if we are given that $X(t) = a$, the distribution of $X(u)$ with $u > t$ is fixed (Cox and Miller, 1977).

The developments of the theoretical understanding of diffusion processes and the derivation of the diffusion equation are historically most interesting. See for instance the review by Narasimhan (1999).

- **Example:** An elastically bound particle undergoing a random walk with a finite step size in one spatial dimension can be described by the model mentioned in Section 7.1. Here the step probabilities were assigned as $\frac{1}{2}(1 + k/R)$ and $\frac{1}{2}(1 - k/R)$ for moving one step to the left or to the right, respectively, provided that the particle is at the position labeled k, see, e.g. Kac (1946). Here, R is a certain integer, and possible positions of the particle are limited by the condition $-R \le k \le R$. The transition probability can consequently be written as $W(k; m) = \frac{1}{2}(1 + k/R)\delta_{m,k-1} + \frac{1}{2}(1 - k/R)\delta_{m,k+1}$ in terms of Kronecker's delta. The generalization of the equation (10.22) for the probability density becomes

$$P(n|m; s+1) = \frac{R+m+1}{2R} P(n|m+1; s) + \frac{R-m+1}{2R} P(n|m-1; s), \tag{10.23}$$

to be solved with the initial condition $P(n|m; 0) = \delta_{m,n}$. Here, $P(n|m; s+1)$ denotes the probability of the particle being at the position with label m at time labeled s, provided that it was at position n initially.

It is readily demonstrated (Wang and Uhlenbeck, 1945) that average values are, for instance,

$$\langle m(s) \rangle = \sum_m m P(n|m; s) = \left(1 - \frac{1}{R}\right) \langle m(s-1) \rangle.$$

Since $\langle m(s) \rangle = n$ at $s = 0$ we have

$$\langle m(s) \rangle = n \left(1 - \frac{1}{R}\right)^s,$$

which shows that the average position of the particle approaches zero. Similarly one finds

$$\langle m^2(s) \rangle = n^2 \left(1 - \frac{2}{R}\right)^s + \frac{R}{2}\left[1 - \left(1 - \frac{2}{R}\right)^s\right],$$

demonstrating that $\langle m^2(s) \rangle \to R/2$ as $s \to \infty$.

In the continuum limit as before, in particular with $R \to \infty$, we find

$$\frac{\partial}{\partial t} P = \gamma \frac{\partial}{\partial x}(xP) + D \frac{\partial^2}{\partial x^2} P, \tag{10.24}$$

with the diffusion coefficient D found before, and $1/(R\tau) \to \gamma$ as $\tau \to 0$. The equation (10.24) is valid for the overdamped case.

- **Example:** As an illustrative example, we consider the dichotomic Markov process. Assume that the particles are moving with uniform, finite velocity $\pm u$ and have a probability α of experiencing a change in velocity per unit time. Denoting the probability densities that a particle moves in the positive and in the negative direction by $p^\pm$, respectively, the master equation for this process becomes

$$\frac{dp_n^+}{dt} = \frac{1-\alpha}{\tau} p_{n-1}^+ + \frac{\alpha}{\tau} p_{n+1}^-, \tag{10.25}$$

$$\frac{dp_n^-}{dt} = \frac{1-\alpha}{\tau} p_{n+1}^- + \frac{\alpha}{\tau} p_{n-1}^+, \tag{10.26}$$

with a self-evident derivation and interpretation. With a Taylor expansion $p_{n\pm1}^\pm \approx p^\pm \pm \Delta x \, dp_n^\pm/dx$ and defining particle probability fluxes and densities as $F = u(p^+ - p^-)$ and $C = (p^+ + p^-)/\Delta x$, respectively, we find an equation for F and C. On eliminating F we find

$$\frac{\partial^2 C}{\partial t^2} + \frac{2\alpha}{\tau} \frac{\partial C}{\partial t} + (2\alpha - 1)u^2 \frac{\partial^2 C}{\partial x^2} = 0. \tag{10.27}$$

When a velocity-change frequency $a \equiv \alpha/\tau$ is introduced, a limiting case can be considered, with $\tau \to 0$, $\Delta x \to 0$, $\alpha \to 0$, $\alpha/\tau = a$, and $\Delta x/\tau = u$. The result is the telegraph equation

$$\frac{\partial^2 C}{\partial t^2} + 2a \frac{\partial C}{\partial t} - u^2 \frac{\partial^2 C}{\partial x^2} = 0. \tag{10.28}$$

The diffusion equation results in the limit $u \to \infty$, $a \to \infty$, in such a way that $u^2/a \to D$. The generalization obtained by describing the classical Brownian motion using the telegraph equation rather than the diffusion equation is immaterial, but (10.28) can be used with advantage for a number of other and different problems. It has been applied to diffusion caused by random, turbulent, motions in fluids, for instance.

The random-walk model may be generalized by introducing a statistical correlation between two successive steps, in such a way that the probability α of a step in the same direction as the previous step differs from the probability β for a step back (a 'random walk with persistence'). In this case the behavior of the position variable, Y, no longer constitutes a Markov process, because its probability density at time t_n depends not only on its value at this time but also on its value at t_{n-1}. However, the Markov property can be restored by introducing the previous position explicitly as an additional variable. The resulting two-component process, (Y_n, Y_{n-1}), is again Markovian, and this model can again be treated in full detail. A stochastic process that can be made Markovian by means of one additional variable is called 'Markovian of second degree,' and, if more variables are needed, it is Markovian of some higher degree (van Kampen, 1981).

- **Example:** Consider the Brownian motion of a simple pendulum in contact with a surrounding gas in thermal equilibrium. The fluctuations in the position of the pendulum can be assumed to be driven by Gaussian random noise, but the displacement $X(t)$ is *not* a Markov process. It is the *projection* of the two-dimensional (Gaussian) Markov process $X(t), p(t)$, where $p(t) \equiv dX(t)/dt$. In this case the correlation function is extended to a correlation matrix

$$||R|| = \left\| \begin{matrix} \langle x(t)x(t+\tau) \rangle & \langle x(t)p(t+\tau) \rangle \\ \langle p(t)x(t+\tau) \rangle & \langle p(t)p(t+\tau) \rangle \end{matrix} \right\| \tag{10.29}$$

  the elements of which are determined by the appropriate Langevin equation (6.19), see Section 6.2.2.

- **Exercise:** Write (10.21) in the symmetric form $P(m; s) = \frac{1}{2}[P(m+1; s+1) + P(m-1; s-1)]$ and demonstrate that the solution is

$$P(n|m, s) = \frac{s!}{[\frac{1}{2}(\nu+s)]![\frac{1}{2}(\nu-s)]!} \left(\frac{1}{2}\right)^s$$

  with $P(m, 0) = \delta_{n,m}$ and $\nu = |n - m|$. See, e.g., Wang and Uhlenbeck (1945).

10.3   Correlation functions and spectra

The master equation, or the Fokker–Planck equation, allows calculation of the correlation function and the fluctuation spectrum for time-stationary processes. It is advantageous to approach the problem via the two-time correlation function defined as $R(\tau) = \langle y(t_1)y(t_1 + \tau)\rangle$, where $y(t)$ is the appropriate variable for the problem. For time-stationary processes we may as well take $t_1 = 0$, i.e. $R(\tau) = \langle y(0)y(\tau)\rangle$. The correlation function is determined by first obtaining the conditional probability density subject to the condition that $y = y_0$ at $t = 0$. The conditional average is then $\langle y, \tau | y_0, 0\rangle = \int_{-\infty}^{\infty} yP(y, \tau | y_0, 0)\, dy$, which is explicitly a function of τ. In the next step we find

$$R(\tau) = \int_{-\infty}^{\infty} \langle y, \tau | y_0, 0\rangle y_0 P(y_0)\, dy_0,$$

where $P(y_0)$ is the equilibrium solution of the probability density. The fluctuation spectrum can then, if desired, be obtained by Fourier transforming $R(\tau)$. By a procedure similar to the one outlined here, the entire two-point probability density can be determined, and so can N-point probability densities.

10.3.1   Correlation functions for Gaussian Markov processes

A particular example of a Markov process is one in which the statistics are Gaussian, in which case the entirety of the statistical information is contained in the correlation function. Because a Markov process on the other hand is completely defined by a joint probability distribution from which the correlation function can be derived, it seems natural to expect that the Markov property imposes some restrictions on $R(t_i, t_j)$. In order to determine the nature of these restrictions we first consider the joint probability density for a Gaussian random process

$$P_2(y_1, t_1; y_2, t_2) = (2\pi)^{-1}|A|^{-1/2} \exp\left(-\frac{1}{2|A|}\sum_{i,j=1}^{2} A_{ij}y_iy_j\right), \tag{10.30}$$

where A_{ij} is the cofactor of

$$R(t_i, t_j) \equiv \langle Y(t_i)Y(t_j)\rangle = \int\int_{-\infty}^{\infty} y_1 y_2 P_2(y_1, t_i; y_2, t_j)\, dy_1\, dy_2$$

in the correlation matrix

$$\|A\| = \left\| \begin{matrix} R(t_1, t_1) & R(t_1, t_2) \\ R(t_2, t_1) & R(t_2, t_2) \end{matrix} \right\|, \tag{10.31}$$

and $|A|$ is the determinant of the matrix, see also Appendix C. Also

$$P_3(y_1, t_1; y_2, t_2; y_3, t_3) = (2\pi)^{-3/2}|B|^{-1/2} \exp\left(-\frac{1}{2|B|}\sum_{i,j=1}^{3} B_{ij}y_iy_j\right), \tag{10.32}$$

where B_{ij} is the cofactor of $R(t_i, t_j)$ in the correlation matrix

$$||B|| = \begin{Vmatrix} R(t_1, t_1) & R(t_1, t_2) & R(t_1, t_3) \\ R(t_2, t_1) & R(t_2, t_2) & R(t_2, t_3) \\ R(t_3, t_1) & R(t_3, t_2) & R(t_3, t_3) \end{Vmatrix},$$ (10.33)

and $|B|$ is the determinant of the matrix. Bayes' rule relates the conditional probability $P(y_3, t_3|y_1, t_1; y_2, t_2)$ to (10.30) and (10.32) by

$$P(y_3, t_3|y_1, t_1; y_2, t_2) = \frac{P_3(y_1, t_1; y_2, t_2; y_3, t_3)}{P_2(y_1, t_1; y_2, t_2)}$$

$$= \sqrt{\frac{|A|}{2\pi|B|}} \exp\left(-\frac{1}{2|B|} \sum_{i,j=1}^{3} B_{ij} y_i y_j + \frac{1}{2|A|} \sum_{i,j=1}^{2} A_{ij} y_i y_j\right).$$ (10.34)

Note that the summation index in the first term in the exponent goes to 3 according to (10.32) but that in the second term goes only to 2 according to (10.30).

In order for $P(y_3, t_3|y_1, t_1; y_2, t_2)$ to be independent of y_1, t_1 as required for a Markov process, it is necessary that

$$B_{13} = 0 = B_{31}$$ (10.35)

and

$$\frac{B_{11}}{A_{11}} = \frac{B_{12}}{A_{12}} = \frac{B_{21}}{A_{21}} = \frac{|B|}{|A|}.$$ (10.36)

The first condition $B_{13} = 0$ implies that

$$R(t_2, t_2)R(t_3, t_1) = R(t_3, t_2)R(t_2, t_1).$$ (10.37)

As yet, no use has been made of the assumption that the signal is time stationary in a statistical sense, implying for instance $R(t_i, t_j) = R(t_i - t_j)$. With this hypothesis, together with normalization of the auto-correlation function so that $R(0) = 1$, which is no loss of generality, the relation (10.37) becomes

$$R(t_3 - t_1) = R(t_3 - t_2)R(t_2 - t_1).$$ (10.38)

This equation, in addition to satisfying $B_{13} = 0$, also makes the various parts of (10.36) valid. The problem now is that of solving the functional relation (10.38). This is done by introducing $\tau = t_3 - t_2$ and $v = t_2 - t_1$. Then $\tau + v = t_3 - t_1$ and (10.38) states that

$$R(\tau + v) = R(\tau)R(v) \qquad \text{for all } \tau, v > 0.$$ (10.39)

Trivial solutions are offered by $R(\tau)$ identically equal to 0 or 1. If $R(v_0) = 0$ at one point v_0, then $R(\tau + v_0) = 0$ for all $\tau > 0$, implying that $R(\tau) = 0$ for $\tau > v_0$. Hence $R(\tau)$ can not equal zero at any point without our obtaining the trivial result that $R(\tau)$ equals zero for all τ to the right of that point. Since, by hypothesis, $R(0) = 1$, this means that $R(\tau) \geq 0$ for all τ. Excluding the equality sign, one has $R(\tau) > 0$ for all τ. It is then legitimate to take the logarithm of both sides of (10.39) since all quantities are positive. Thus

$$\ln R(\tau + v) = \ln R(\tau) + \ln R(v).$$ (10.40)

If $R(\tau)$ is analytic at $\tau = 0$, that is, has a power-series expansion about $\tau = 0$, the same is true of $\ln R(\tau)$. Assume that

$$\ln R(\tau) = \sum_{n=0}^{\infty} c_n \tau^n, \tag{10.41}$$

where the coefficients c_n are to be determined. Then, from (10.40)

$$\sum_0^{\infty} c_n(\tau + v)^n = \sum_0^{\infty} c_n \tau^n + \sum_0^{\infty} c_n v^n. \tag{10.42}$$

If it is satisfied at all, this relation is satisfied for all (τ, v), provided that $c_n = 0$ for $n = 0$ and for all $n \geq 2$. Hence $\ln R(\tau)$ must be

$$\ln R(\tau) = c_1 \tau. \tag{10.43}$$

It is known *a priori* that $R(\tau)$ is an even function of τ that approaches zero as $\tau \to \infty$. This implies that the coefficient c_1 is negative. Therefore, letting $c_1 = -\beta$, where $\beta > 0$, the final result is

$$\ln R(\tau) = -\beta|\tau|.$$

Solving for $R(\tau)$, we find

$$R(\tau) = e^{-\beta|\tau|}, \qquad \text{with } \beta > 0. \tag{10.44}$$

This completes the desired proof (Bendat, 1958) and also demonstrates the fundamental importance of exponential auto-correlation functions. This theorem provides a strong theoretical justification for expecting many random-noise phenomena to exhibit exponential auto-correlation functions.

The spectrum for a Gaussian Markov process is obtained by Fourier transforming the correlation function (the Wiener–Khinchine theorem) and it is found to have the universal, Lorentzian, form

$$S(\omega) = \frac{2\beta}{\beta^2 + \omega^2}. \tag{10.45}$$

The micro timescale defined as $[-d^2 R(\tau)/d\tau^2|_{\tau=0}]^{-1/2}$ vanishes for correlation functions of the form (10.44). This is consistent with the Markov property, i.e. the lack of memory exhibited by the process. Correspondingly, we have a divergence in the derivative process, $\langle (dY(t)/dt)^2 \rangle = \int \omega^2 S(\omega) \, d\omega \to \infty$, indicating that Gaussian Markov processes usually have singularities in their time derivatives.

For modeling actual spectra, the result obtained here is often considered too restrictive. Deviations from Gaussianity are allowed for and correlation functions of the form $R(\tau) = Ae^{-\beta|\tau|} \cos(\Omega t)$ are assumed, wherein the constants β and Ω are obtained by fits to results based on actual data. The corresponding frequency spectrum becomes

$$S(\omega) = \frac{2A\beta}{\pi} \left(\frac{\omega^2 + \beta^2 + \Omega^2}{\omega^4 + 2(\beta^2 - \Omega^2)\omega^2 + (\beta^2 + \Omega^2)^2} \right). \tag{10.46}$$

When $3\Omega^2 > \beta^2$, this spectrum has a peak for $\omega_1 = (\beta^2 + \Omega^2)^{1/4}[2\Omega - (\beta^2 + \Omega^2)^{1/2}]^{1/2}$ with a peak power $S(\omega_1) = (2A\beta/\pi)/(\Omega^2 + \beta^2)$. Surprisingly many experimentally obtained power spectra can be approximated by (10.46) with proper choices of Ω and β.

- **Example:** As an important special case we consider an amplitude- and phase-modulated harmonic oscillation, for which one record, or one realization, labeled k is given as

$$^k s(t) = {}^k a(t) \sin(\Omega t + {}^k \theta),$$

where Ω is constant, while a and θ are mutually independent random variables, where in particular θ is uniformly distributed over the interval $\{0; 2\pi\}$. Suppose that the auto-correlation of a is given by $R_{aa}(\tau) \equiv \langle a(t)a(t+\tau) \rangle = A \exp(-\beta|\tau|)$. The auto-correlation of $s(t)$ is then readily obtained as

$$R_{ss}(\tau) = \langle a(t) \sin(\Omega t + \theta(t)) \, a(t+\tau) \sin(\Omega(t+\tau) + \theta) \rangle$$
$$= R_{aa}(\tau) \frac{1}{2\pi} \int_0^{2\pi} \sin(\Omega t + \theta) \sin(\Omega(t+\tau) + \theta) \, d\theta$$
$$= \tfrac{1}{2} A e^{-\beta|\tau|} \cos(\Omega \tau),$$

demonstrating a possible way of obtaining an exponential-cosine correlation function.

11 Diffusion of particles

The Fokker–Planck equations, as well as the diffusion equations, discussed in Chapter 10 and also briefly in Section 6.1.1 are dynamic equations for the space–time variation of *probability densities*. This is in principle a problem that is entirely different than the description of the spreading of a drop of colored fluid in a glass of water, although we expect that also *this* problem can be described by a diffusion equation, at least within certain limits. Here, we briefly outline the derivation of diffusion equations for this case, and consider a fluid with mass density $\rho \equiv Mn$, where n is the *particle* density, a fluid velocity u, pressure p, dispersing in a background medium, which interacts with the fluid through an effective friction coefficient v.

The starting point is here the fluid equations, i.e. the continuity equation

$$\frac{\partial}{\partial t} n(r, t) + \nabla \cdot [n(r, t)u(r, t)] = 0, \tag{11.1}$$

where $n(r, t)u(r, t)$ is the flux density of fluid material at position r at time t. As the second equation we have the momentum equation in the form

$$Mn(r, t)\frac{D}{Dt}u(r, t) \equiv$$
$$Mn(r, t)\left(\frac{\partial}{\partial t}u(r, t) + u(r, t) \cdot \nabla u(r, t)\right) =$$
$$- \nabla p(r, t) - vMn(r, t)[u(r, t) - U_b(r, t)], \tag{11.2}$$

where D/Dt is the *convective* or *material* time derivative, $\partial/\partial t$ is the *local* or Eulerian time derivative, and $U_b(r, t)$ is the velocity of the background flow. The thermal agitation of the particles in the drop of colored fluid is accounted for by the finite temperature entering into the pressure, while the convection due to the surrounding flow is given through $U_b(r, t)$. Now assume that $U_b(r, t) = 0$ and that the relevant processes are so slow that $Du(r, t)/Dt \ll vu(r, t)$. (In reality this is a rather strict assumption; molecular diffusion is a very slow and in a sense 'ineffective' process, and extreme care must be taken to ensure that the background fluid is indeed at rest.) In this case (11.2) can be simplified to $\nabla p(r, t) = -vn(r, t)u(r, t)$. Consistent with the assumption of a slow variation with time we let $p \approx n\kappa T$ for an isothermal process, for which the fluid specified by (11.1) and (11.2) is assumed to be in thermal equilibrium with the surrounding fluid at temperature T. Using this expression for p, the result is readily inserted into (11.1) to give a diffusion equation with diffusion coefficient $D = \kappa T/(Mv)$, see also (6.5).

If, for some reason, the friction is spatially varying, $v = v(r)$, because of variations in the background flow density, for instance, we have $D = D(r)$. The diffusion equation is then of the form

$$\frac{\partial}{\partial t} n(\mathbf{r}, t) = \nabla \cdot [D(\mathbf{r}) \nabla n(\mathbf{r}, t)].\tag{11.3}$$

By comparison with the equation of continuity (11.1), we see that the diffusive particle flux density is $-D\nabla n$; this is Fick's law.

It is important to emphasize that, although the diffusion equation for the particle density discussed in this section is formally the same as the diffusion equation for the probability density, the underlying physics is completely different. When discussing the space–time evolution of the probability density, we imagine many realizations (ideally, infinitely many) of the same experiment whereby *one* particle is released at the origin, and its position followed in time. From these many realizations, the probability density for the position of the particle at time t is eventually estimated, and the analytic result predicts that this probability density follows a diffusion equation. When discussing diffusion of one blob of ink, as we are here, only one realization is addressed. If the individual particles, atoms or molecules, in the ink are bombarded *independently* by the particles in the surrounding fluid, we can argue that one physical realization corresponds to an entire ensemble of realizations insofar as the individual particles in the ink are concerned. Evidently, this interpretation is based on somewhat optimistic assumptions. For physical conditions under which the hypothesis can be justified, it does not really matter whether we consider particle densities or probability densities; for those conditions the two concepts are freely interchangeable. The density of particles originating from a point source at $t = 0$ is then trivially related to the probability density by $n(\mathbf{r}, t) = NP(\mathbf{r}, t)$, where N is the number of particles released at the source at $t = 0$. Again, the probability density can here be interpreted as the ratio of desired observations to the total number of relevant observations, here N, i.e. $P(\mathbf{r}, t)\, d\mathbf{r} = n(\mathbf{r}, t)\, d\mathbf{r}/N$.

A straightforward generalization of (11.2) has interesting applications. For this case we assume that an external force is included in (11.2) to give

$$Mn(\mathbf{r}, t) \frac{D}{Dt} \mathbf{u}(\mathbf{r}, t) \equiv Mn(\mathbf{r}, t) \left(\frac{\partial}{\partial t} \mathbf{u}(\mathbf{r}, t) + \mathbf{u}(\mathbf{r}, t) \cdot \nabla \mathbf{u}(\mathbf{r}, t) \right) =$$
$$- \nabla p(\mathbf{r}, t) - \nu Mn(\mathbf{r}, t)[\mathbf{u}(\mathbf{r}, t) - \mathbf{U}_b(\mathbf{r}, t)] + Mn(\mathbf{r}, t)\mathbf{K}(\mathbf{r}),\tag{11.4}$$

where $\mathbf{K}$ is an acceleration caused by an external force field. Generally, we expect $\mathbf{K}$ to be spatially varying. As before, we argue that, for slow processes with $\mathbf{U}_b(\mathbf{r}, t) = 0$, the inertia term on the left-hand side of the equation can be ignored. In this limit we find that $-\nabla p(\mathbf{r}, t) - \nu Mn(\mathbf{r}, t)\mathbf{u}(\mathbf{r}, t) - Mn(\mathbf{r}, t)\mathbf{K}(\mathbf{r}) \approx \mathbf{0}$, giving a generalization of the diffusion equation of the form

$$\frac{\partial}{\partial t} n(\mathbf{r}, t) = \nabla \cdot \left(D \nabla n(\mathbf{r}, t) - \frac{\mathbf{K}(\mathbf{r})}{\nu} n(\mathbf{r}, t) \right).\tag{11.5}$$

This equation is often called *Smoluchowski's equation* (not to be confused with (10.9)) when it is expressed in terms of probability densities. The case $\mathbf{K} = -\alpha x \hat{\mathbf{x}}$ in (11.5) corresponds to a harmonic oscillator with $\hat{\mathbf{x}}$ being a unit vector.

For stationary conditions the diffusion flux density is constant, i.e.,

$$\mathbf{j} \equiv D \nabla n(\mathbf{r}, t) - \frac{\mathbf{K}(\mathbf{r})}{\nu} n(\mathbf{r}, t) = \text{constant}.\tag{11.6}$$

When $\mathbf{K}$ can be derived from a potential, $\mathbf{K} = -\nabla \Phi(\mathbf{r})$, we can rewrite (11.6)

$$j = D \exp[-\Phi/(\nu D)] \, \nabla\{n \exp[\Phi/(\nu D)]\} = \text{constant}, \tag{11.7}$$

where we recall that $D\nu = \kappa T/m$ according to the Einstein relation. Integrating (11.7) between any two points A and B, we have under stationary conditions

$$j \cdot \int_A^B \nu \exp[\Phi/(\nu D)] \, ds = \frac{\kappa T}{m} n \exp[\Phi/(\nu D)] \Big|_A^B, \tag{11.8}$$

usually called Kramer's relation, which is important for analyzing diffusion across potential barriers, see, e.g., Chandrasekhar (1943).

- **Exercise:** Solve the diffusion equation $\partial n/\partial t = D \, \nabla^2 n$ in cylindrical and in spherical geometries for the case in which the initial condition is $n(r, 0) = \delta(|r| - R_0)$, where R_0 is a constant.

- **Exercise:** Consider a point source from which particles are released at a constant rate starting at time $t = 0$, at a distance ℓ from a fixed plane wall. Assume that the dispersion of the released particles is described by a simple diffusion equation. Use the method of images to obtain the space–time-varying density distribution of particles for the cases in which the wall is (i) absorbing and (ii) reflecting.

- **Exercise:** Discuss diffusion in a linear shear flow $V \equiv \{V_x, V_y, V_z\} = \{\alpha y, 0, 0\}$, assuming that particles disperse solely by classical diffusion in the y- and z-directions, with diffusion coefficient D. Demonstrate that $\langle Y^2 \rangle = Dt$, $\langle Z^2 \rangle = Dt$, and $\langle X^2 \rangle = \frac{1}{3}\alpha^2 Dt^3$.

- **Exercise:** Solve the diffusion equation $\partial n/\partial t = D \, \nabla^2 n$ in cylindrical and in spherical geometries for the case in which the initial condition is $n(r, 0) = n_0 H(|r| - R_0)$, where H is Heaviside's step function, and R_0 is a constant. Assume that the boundary at $|r| = R_0$ is absorbing, and obtain an expression for the flux density $-D \, \nabla n$ through the sphere with radius R_0 for the spherical case, and for a cylinder of radius R_0 and unit length in the z-direction in the cylindrically symmetric case. For the cylindrical case you will need the identity

$$1 = \frac{2}{\pi} \int_0^\infty \frac{J_0(\lambda R) Y_0(\lambda r) - Y_0(\lambda R) J_0(\lambda r)}{J_0^2(\lambda R) + Y_0^2(\lambda R)} \frac{d\lambda}{\lambda},$$

in terms of the Bessel functions J_0 and Y_0 of first and second kinds, respectively (Abramowitz and Stegun, 1972).

- **Exercise:** Demonstrate that a solution of the diffusion equation $\partial n/\partial t = D(t) \, \nabla^2 n$ with initial condition $n(r, 0) = \delta(|r|)$ and a time-varying diffusion coefficient $D(t)$ is

$$n(r, t) = \left(4\pi \int_0^t D(t') \, dt' \right)^{-3/2} \exp\left(-r^2 \Big/ 4 \int_0^t D(t') \, dt' \right)$$

for any nonsingular $D(t)$.

For the case in which K depends on one coordinate only, the relation (11.5) has simple stationary solutions, $\partial n(x)/\partial t = 0$,

$$n(x) = n_0 \exp\left(\int_{-\infty}^{x} K(\xi)/(vD)\,d\xi\right). \tag{11.9}$$

The relation (8.3) used in Chapter 8 is just a special case of (11.9).

11.1   Applications

As actual examples for application of the Fokker–Planck equation, or a generalization of the diffusion equation, we here consider two problems: (i) sedimentation in isothermal atmospheres in gravitational fields; and (ii) coagulation in colloids.

11.1.1   Sedimentation in atmospheres

For the present problem we use Smoluchowski's equation (11.5) with K being a constant gravitational acceleration. Accounting also for the updrift caused by the surrounding medium (Archimedes' law), we have $K = \{0, 0, -(1 - \rho_0/\rho)g\}$, where ρ_0 is the mass density of the surrounding fluid, $\rho > \rho_0$ is the density of the particles, and g is the gravitational acceleration. The problem can be formulated in one spatial dimension without loss of generality, and we find the basic equation

$$\frac{\partial}{\partial t}n(r, t) = D\frac{\partial^2}{\partial x^2}n(r, t) + C\frac{\partial}{\partial x}n(r, t). \tag{11.10}$$

The equation was simplified by introducing $C = (1 - \rho_0/\rho)(g/v)$.

Equation (11.10) will be solved in one spatial dimension with the initial condition $n(x, t = 0) = N\delta(x - x_0)$, together with the boundary condition that the net particle flux at $x = 0$ vanishes at all times. This requirement can be satisfied by the condition $D\,\partial n(x, t)/\partial x|_{x=0} + Cn(x = 0, t) = 0$, for all $t > 0$. This boundary condition corresponds to a perfectly reflecting ground at $x = 0$, see for instance Section 7.1.4. The dimension of n here is reciprocal length. It is advantageous to introduce the transformation

$$n(x, t)/N = U(x, t)\exp[-C(x - x_0)/(2D) - tC^2/(4D)],$$

and obtain the standard form $\partial U(x, t)/\partial t = D\,\partial^2 U(x, t)/\partial x^2$, while the boundary condition becomes $D\,\partial U(x, t)/\partial x + \frac{1}{2}CU(x, t) = 0$. The solution of the diffusion equation with this boundary condition is

$$U(x, t) = \frac{1}{2\sqrt{\pi Dt}}\left[\exp\left(-\frac{(x - x_0)^2}{4Dt}\right) + \exp\left(-\frac{(x + x_0)^2}{4Dt}\right)\right]$$

$$+ \frac{C}{2D\sqrt{\pi Dt}}\int_{x_0}^{\infty} \exp\left(-\frac{(\xi + x)^2}{4Dt} + \frac{C(\xi - x_0)}{2D}\right)d\xi. \tag{11.11}$$

After some simple transformations this result is rewritten as

$$U(x, t) = \frac{1}{2\sqrt{\pi Dt}} \left[\exp\left(-\frac{(x - x_0)^2}{4Dt} \right) + \exp\left(-\frac{(x + x_0)^2}{4Dt} \right) \right]$$

$$+ \frac{C}{D\sqrt{\pi}} \exp\left(\frac{C^2 t}{4D} - \frac{C(x + x_0)}{2D} \right) \int_{\frac{x+x_0-Ct}{2\sqrt{Dt}}}^{\infty} \exp(-\xi^2) \, d\xi. \tag{11.12}$$

This result can be re-expressed in terms of the original variable $n(x, t)$ as

$$\frac{n(x, t)}{N} = \frac{1}{2\sqrt{\pi Dt}} \left[\exp\left(-\frac{(x - x_0)^2}{4Dt} \right) + \exp\left(-\frac{(x + x_0)^2}{4Dt} \right) \right]$$

$$\times \exp\left(-\frac{C^2 t}{4D} - \frac{C(x - x_0)}{2D} \right)$$

$$+ \frac{C}{D\sqrt{\pi}} \exp\left(-\frac{Cx}{D} \right) \int_{\frac{x+x_0-Ct}{2\sqrt{Dt}}}^{\infty} \exp(-\xi^2) \, d\xi. \tag{11.13}$$

It is evidently advantageous to introduce the normalized variables $\tau = tC^2/D$ and $y = xC/D$. Then only y_0 remains as a parameter for the problem. The solution is shown as a function of y for various times in Fig. 11.1, for $y_0 = 2.0$. In this case, the particles first begin to disperse due to diffusion, while they simultaneously fall to the ground. When they reach ground level at $y = 0$ they bounce back, and eventually relax to a stationary distribution. The integral of (11.13) over all x is conserved. In the case in which the particles are released at ground level, shown in Fig. 11.2 with $y_0 = 0$, the particles spread by diffusion and again attain a stationary distribution with altitude. In both cases, the ultimate stationary solution is

$$\frac{n(x)}{N} = \frac{C}{D} \exp\left(-\frac{Cx}{D} \right),$$

where $D/C \equiv [\kappa T/(Mg)]/(1 - \rho_0/\rho)$ is the scale height for an isothermal atmosphere.

For the analysis in this section it does not really matter whether we consider particle densities or probability densities. The concepts are freely interchangeable for the present problem.

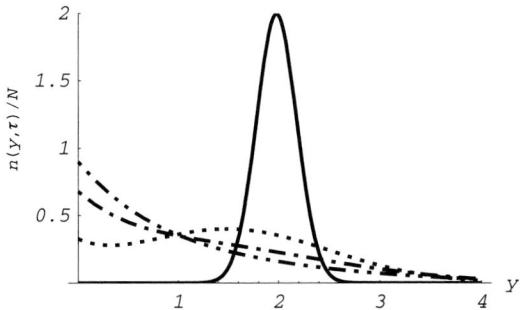

Figure 11.1 An illustration of sedimentation of particles in a gravitational field, for the case in which particles are released at $y_0 = 2.0$. The normalized variables used in the figure are defined in the text. The vertical axis shows the normalized density. Times are indicated by the full line, $\tau = 0.02$; dashed line, $\tau = 0.5$; dot–dashed line $\tau = 1.0$; and dashed–double-dotted line, $\tau = 2.0$.

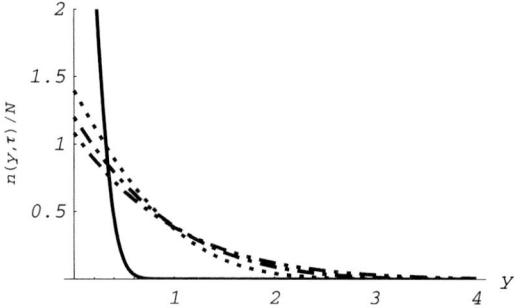

Figure 11.2 An illustration of sedimentation of particles in a gravitational field, for the case in which particles are released at $y_0 = 0$. The normalized variables used in the figure are defined in the text. The vertical axis shows the normalized density. Times are indicated by the full line, $\tau = 0.02$; dashed line, $\tau = 0.5$; dot–dashed line, $\tau = 1.0$; and dashed–double-dotted line, $\tau = 2.0$.

- **Exercise:** How would you formulate the problem of sedimentation for the case in which the particles are *lighter* than the surrounding fluid, i.e. $\rho_0 > \rho$?

11.1.2　Coagulation in colloids

The analysis of Brownian motion presented hitherto assumes that the parameters relevant for the problem (e.g. diffusion coefficients) remain constant. This need not be so, and as an illustration of such a nonstationary problem we discuss here coagulation in colloids. The problem was seemingly first studied by Smoluchowski, but we here follow the exposition of Chandrasekhar (1943).

The discussion here concerns the temporal evolution of a system containing macroscopic particles that are allowed to coalesce, thereby forming new particles of a different size. The diffusion coefficient of particles depends on size according to the formula (6.18), and the diffusion of the newly formed particles progresses at a different rate. Assuming that the time between coagulation of particles is long compared with the short timescale for the bombardment of the large particles by the atoms or molecules in the surrounding fluid, it can be argued that the diffusion of the large particles is at all times controlled by a diffusion coefficient given by (6.18), i.e. the process is at all times in local thermal equilibrium. This is a very modest restriction for relevant processes.

Physically, we assume that we have a dilute solution of many particles, which diffuse independently by Brownian motion, with diffusion coefficient $\kappa T/(6\pi\eta a)$, where a is the diameter of the particle, see (6.18). When incidentally two such particles approach each other sufficiently closely that their separation is less than the radius of a certain *sphere of influence*, they are assumed to stick together with probability unity, and diffuse as one particle from then on, with a new diameter, however, and therefore a new diffusion coefficient. The probability of *three* particles coalescing simultaneously is considered negligible. This assumption implies that the radius of the sphere of influence is much smaller than the average distance between colloid particles.

First we find the result for the rate of formation of multiple particles by coagulation. The starting point of the analysis is the simple diffusion equation. First assume that we have a reference particle at rest and placed in an initially uniform density of surrounding particles. We determine the probable rate of arrival of other particles coming within its sphere of influence, with radius R assumed given. When the particles reach this distance from the reference particle, they are lost in the sense that they enter as constituents into the concentrations of new, larger, particles. We therefore solve $\partial n/\partial t = D\,\nabla^2 n$ with the boundary condition $n = 0$ for $|r| = R$ at all times just like for an absorbing barrier, while the initial condition is $n = n_0 = $ constant for all $r > R$, at $t = 0$. Owing to the spherical symmetry of the problem, we can rewrite the diffusion equation for the particle density as

$$\frac{\partial}{\partial t}[r\,n(r,t)] = D\,\frac{\partial^2}{\partial r^2}[r\,n(r,t)], \tag{11.14}$$

where the radius, r, and time, t, are the only independent variables. The solution of this equation for $r > R$, with the given boundary conditions, is

$$n(r,t) = n_0\left(1 - \frac{R}{r} + \frac{2R}{r\sqrt{\pi}}\int_0^{(r-R)/(2\sqrt{Dt})} \exp(-\xi^2)\,d\xi\right). \tag{11.15}$$

The particle flux through a sphere with radius r is then $4\pi r^2 D\,\partial n/\partial r$, here to be obtained at $r = R$. We find that

$$4\pi r^2 D\,\frac{\partial}{\partial r}n(r,t)\bigg|_{r=R} = 4\pi DRn_0\left(1 + \frac{R}{\sqrt{\pi Dt}}\right), \tag{11.16}$$

in units of number of particles per unit time, with n_0 being the initial density. This will also be the constant density level at large distances from the reference particle. Physically, (11.16) gives a particle flux as an average over many independent reference particles. With the initial assumption relevant for the present problem, this particle flux will be small. If the density n_0 is allowed to be large, i.e. many particles enter the sphere of influence within a small time interval, then the particle flux given by (11.16) will be approximately the one experienced by *one* reference particle.

The result (11.16) accounts for the rate at which macroscopic particles of a given type (with a given diffusion coefficient) coalesce with a *stationary* macroscopic particle surrounded by a sphere of influence. This reference particle is, however, itself participating in the Brownian motion. Its motion is also caused by the impact of atoms and molecules of the surrounding fluid, and it can be considered independent of the motion of all the other macroscopic particles. The *relative displacement* $r_{1,2}$ of two independently diffusing particles with positions r_1 and r_2 is obtained by noting that $d\langle r_{1,2}^2\rangle/dt \equiv d\langle (r_1 - r_2)^2\rangle/dt = d\langle r_1^2\rangle/dt + d\langle r_2^2\rangle/dt$, with the motions of the two particles being independent, implying $d\langle r_1 \cdot r_2\rangle/dt = 0$, with $\langle r_1\rangle$ and $\langle r_2\rangle$ being constants. With the assumption of diffusive motion we have $\langle r_1^2\rangle = 2D_1 t$ and $\langle r_2^2\rangle = 2D_2 t$, see (6.17). Assuming that the variation of $r_{1,2}$ is also diffusive, we readily find that $\langle r_{1,2}^2\rangle = 2D_{1,2}\,t$ with $D_{1,2} \equiv D_1 + D_2$.

We can now generalize (11.16) to

$$4\pi r^2(D_1 + D_2)\,\frac{\partial}{\partial r}n(r,t)\bigg|_{r=R} =$$
$$4\pi(D_1 + D_2)Rn_0\left(1 + R/\sqrt{\pi(D_1 + D_2)t}\right). \tag{11.17}$$

Introduce now $R_{i,j}$ as the radius of the sphere of influence for the encounter of two particles of different sizes, labeled by i and j. In terms of the relative diffusion coefficients $D_{i,j}$, the evident generalization of (11.16) for the particle flux $J_{i,j}$ in a small time interval dt then becomes

$$J_{i,j}\, dt = 4\pi D_{i,j} R_{i,j} n_{0i} n_{0j} \left(1 + \frac{R_{i,j}}{\sqrt{\pi D_{i,j} t}}\right) dt, \qquad (11.18)$$

where n_{0j} is the initial density of particles of type j at $r > R_{i,j}$. The singularity at $t = 0$ is due to the unrealistic assumption of an initial step function in the particle density at $r = R_{i,j}$. For large times, $t \gg R_{i,j}^2/(\pi D)$, we can ignore the second term in (11.18) and find that

$$J_{i,j}\, dt \approx 4\pi D_{i,j} R_{i,j} n_{0i} n_{0j}\, dt. \qquad (11.19)$$

For practical cases, the time $R_{i,j}^2/D$ can be estimated as 10^{-4}–10^{-3} s, so this restriction is of little consequence.

The relation (11.19) describes the basic process for the problem of coagulation as an initial value problem, with densities given at $t = 0$. For the dynamic situation with time-varying concentrations that is relevant here, it is trivially argued that the rate at which particles of type j are transformed into other types is proportional to the instantaneous concentration $\eta_j(t)$. However, also the concentration of the surrounding particles is changing. To account for this effect, consider the diffusion equation with a spatially uniform 'sink' $f(t)$, which depletes particles uniformly in space, i.e. $\partial n/\partial t - D \nabla^2 n = f(t)n$. By introducing $n(r, t) = g(r, t)\exp(\int^t f(t')\, dt')$, it is easily shown that, after a transient time, the flux of particles through a reference sphere with radius R is $4\pi D R \eta(t)$, where $\eta(t)$ is the instantaneous concentration of the particles at time t. Consequently, the rate at which the reference, j-type, particle is 'bombarded' at any time by i-type particles is proportional to $\eta_i(t)$. Therefore, n_{0i} and n_{0j} in (11.19) can be replaced by $\eta_i(t)$ and $\eta_j(t)$ to obtain the instantaneous flux $J_{i,j}(t)$.

- **Exercise:** Consider two independent particles diffusing randomly with diffusion coefficients D_1 and D_2, both having Gaussian probability densities for the particle's displacement from a common origin from which the particle is released. Demonstrate that the probability density of the interparticle separation is also a Gaussian. Show that the separation process is also diffusive, with a diffusion coefficient $D_{1,2} = D_1 + D_2$. Derive an expression for the mean-square average particle position $\langle\frac{1}{4}(r_1 + r_2)^2\rangle$.

Assume that initially all the particles are of the same size, and consider a process that has continued for some time. Irrespective of the initial condition, there will at late times be particles that are composed of single, double, etc. coagulates, with concentrations $\eta_1, \eta_2, \ldots$. We will from now on consider the time-varying concentration $\eta_k = \eta_k(t)$ of k-fold coagulants, i.e. the number of particles divided by the volume of the system, as distinct from the space–time-varying particle density $n_k(r, t)$ in, e.g., (11.14). A gain–loss equation is readily obtained as

$$\frac{d}{dt}\eta_k(t) = \frac{1}{2}\sum_{i+j=k} J_{i,j} - \sum_{j=1} J_{k,j}. \qquad (11.20)$$

The first summation accounts for the increase of η_k due to coagulation of i-fold and j-fold particles, which have $i + j = k$, where the factor $\frac{1}{2}$ compensates for double counting. The second sum in (11.20) accounts for the loss of k-fold particles due to formation of $(k + j)$-fold new particles with a summation over all j. With the approximation mentioned before, we can write (11.20) as

$$\frac{d}{dt}\eta_k(t) = 4\pi \left(\frac{1}{2} \sum_{i+j=k} D_{i,j} R_{i,j} \eta_i \eta_j - \eta_k \sum_{j=1} D_{k,j} R_{k,j} \eta_j \right), \tag{11.21}$$

for $k = 1, 2, \ldots$.

A general solution of (11.21) is rather hopeless, but its properties can be illustrated by a model based on simplifying assumptions. First, it will be assumed that the diameter of the new particle is the average of the diameters of the two particles from which it is composed, $R_{i,j} = \frac{1}{2}(R_i + R_j)$. This assumption can evidently be criticized, but there will be so many other assumptions that this particular simplification does not severely restrict the result. With reference to the basic expression $D = \kappa T/(6\pi\eta a)$ for the diffusion coefficient, we can also argue that $D_i R_i = DR$, where DR is a constant. With these two basic expressions, we find $D_{i,j} R_{i,j} = \frac{1}{2}(R_i + R_j)(D_i + D_j) = \frac{1}{2}DR(R_j + R_i)^2/(R_i R_j)$. Finally, for the sake of mathematical simplicity, we assume that $R_i = R_j$, which is probably the weakest point in the present argument. All in all we find $D_{i,j} R_{i,j} = 2DR$. The basic equation (11.21) can then be rewritten as

$$\frac{d}{dt}\eta_k(t) = 8\pi DR \left(\frac{1}{2} \sum_{i+j=k} \eta_i \eta_j - \eta_k \sum_{j=1} \eta_j \right), \qquad k = 1, 2, \ldots. \tag{11.22}$$

For a finite volume, η_k will be statistically varying over the ensemble of realizations. We assume here that the volume is so large that the variations in η_k from one experiment to the next are negligible.

By introducing a new variable $\tau = 4\pi DR\,t$, having dimension length^3, we can write (11.22) in a more convenient form as

$$\frac{d}{d\tau}\eta_k(t) = \sum_{i+j=k} \eta_i \eta_j - 2\,\eta_k \sum_{j=1} \eta_j, \qquad k = 1, 2, \ldots. \tag{11.23}$$

From (11.23) we readily obtain by summation

$$\frac{d}{d\tau}\sum_{k=1} \eta_k(t) = \sum_{j=1}\sum_{i=1} \eta_i \eta_j - 2\sum_{k=1}\eta_k \sum_{j=1} \eta_j$$

$$= -\left(\sum_{k=1} \eta_k\right)^2, \tag{11.24}$$

giving

$$\sum_{k=1} \eta_k(t) = \frac{\eta_0}{1 + \eta_0 \tau}, \tag{11.25}$$

where we have introduced $\sum_k \eta_k \equiv \eta_0$ at $\tau = 0$. The total number of colloidal particles is thus steadily decreasing. The vanishing asymptotic concentration implies that ultimately only one particle remains in a large volume.

Using (11.25), we can successively obtain solutions for η_1, η_2, etc., and find for instance $d\eta_1/d\tau = -2\eta_1 \sum_{k=1} \eta_k = -2\eta_1\eta_0/(1 + \tau\eta_0)$, or $\eta_1 = \eta_0/(1 + \tau\eta_0)^2$, using $\eta_1 = \eta_0$ at $\tau = 0$. By induction the general result is obtained as

$$\eta_k(t) = \eta_0 \frac{(\eta_0\tau)^{k-1}}{(1+\eta_0\tau)^{k+1}}, \qquad k = 1, 2, \dots \tag{11.26}$$

which is illustrated in Fig. 11.3 for $k = 1, 2, 3$, and 4, shown with full, dashed, dot–dashed and dashed–double-dotted lines, respectively. In Fig. 11.3 the normalized variables η_j/η_0 and $t = \eta_0\tau$ are used. Evidently, the density of the initial particle population decreases steadily, while that for the higher order coagulation products increases for a little while, until also these populations are depleted by coagulation leading to the formation of other particles. Asymptotically we have

$$\frac{\eta_k}{\eta_0} \to \frac{1}{(\eta_0\tau)^2},$$

independent of k.

In spite of the simplifying assumptions made in the analysis, the results seem to reproduce the basic properties of the process quite well. In the limit $t \to \infty$, the analysis predicts that only one giant particle remains; in reality of course very large coagulants will be fragile and break up. This process will enter as a loss for large values of k in (11.21) and a gain for smaller values of k. Initially, $d\eta_k(t)/dt = 0$ for all $k > 2$, while $\eta_2(t)$ increases and $\eta_1(t)$ decreases linearly with t for short times; initially, in a short time interval, we have no production of triple or higher order coagulants since we ignored the possibility of two or more particles coalescing simultaneously.

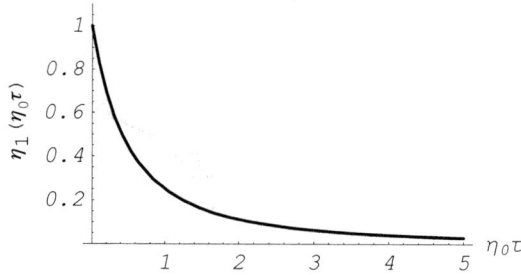

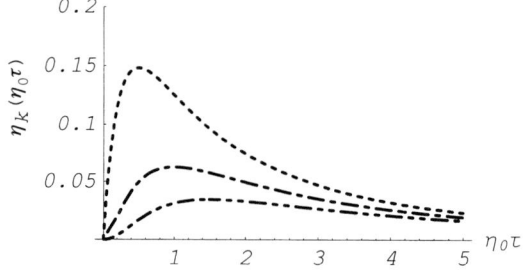

Figure 11.3 An illustration of coagulation of particles, for the case in which all particles are of the same size initially. For the full line we have $k = 1$, for the dashed line $k = 2$, for the dot–dashed line $k = 3$, and for the dashed–double-dotted line $k = 4$. Note the change in scale between the figures.

In spite of the limitations regarding the region of validity, the basic result might seem to pose a puzzle; it assumes that there is an initially disorganized medium containing small particles that are randomly distributed. As time passes, the system contains fewer and fewer, but larger, particles, implying that a sort of self-organization is occurring, in spite of the fact that the process was assumed to proceed in thermal equilibrium at a given temperature T. At first sight, such a process is in contradiction to a variant of the Second Law of Thermodynamics. In reality, the analysis is evidently restrictive in that it considers only particle densities, and not at all the processes occurring *in* the particles themselves, in particular not the chemical processes involved when they stick together. Self-organization is found to occur by considering the *restricted information* contained in the number of particles.

12   Thermal fluctuations in a diode

Fluctuations in a circuit containing a diode were discussed briefly in Section 4.1. It was argued that a physical diode in thermal equilibrium is itself contributing to the fluctuations. For very small fluctuation levels such that the diode characteristic may be linearized, the circuit can be considered as being composed of simple linear components. The contribution from the non-linearities can, however, be interesting and will be discussed here by reference to a specific example. Since the diode is itself contributing to the fluctuations, no additional resistive fluc-tuating element is necessary for the arguments. The circuit discussed here therefore contains only a diode and a capacitor. The question is that of whether the nonlinearity of the diode characteristic will give rise to a net building up of charge on the capacitor plates, i.e. whether the diode can rectify its own fluctuations! This particular problem was analyzed by, for instance, Alkemade (1958) and later in more detail by van Kampen (1960, 1961, 1963).

12.1   The master equation

Consider the circuit shown in Fig. 12.1 containing a diode and a capacitor, with the diode consisting of two parallel plane electrodes close to each other. The two halves of the circuit, including the capacitor plates, are assumed to be made of two different metals with work functions W_1 and W_2. The space between the two capacitor plates is filled with some insulating dielectric, which inhibits motion of charged particles from one capacitor plate to the other.

Suppose that electrons are emitted at a constant rate from electrode 1 according to Richardson's formula. The probability of an electron leaving in a time interval of duration dt is then

$$P_1 \, dt = \frac{4\pi m}{h^3} (\kappa T)^2 \mathcal{A} e^{-W_1/(\kappa T)} dt \equiv \xi \, dt, \tag{12.1}$$

where $\mathcal{A}$ is the area of the electrode (Pathria, 1996). Electrons emitted from electrode 2 can reach electrode 1 only if they have sufficient kinetic energy to overcome the potential difference. Let there be an imbalance between the numbers of electrons in the two halves of the circuit with an excess of N electrons. The voltage across the capacitor is then eN/C. The electrostatic energy of the system is $e^2 N^2/(2C)$, where the capacitance of the diode itself is considered negligible. An electron can leave electrode 2 and reach electrode 1 only if its kinetic energy is larger than

$$\frac{e^2}{2C}(N+1)^2 - \frac{e^2}{2C}N^2 = \frac{e^2}{C}\left(N+\frac{1}{2}\right). \tag{12.2}$$

The probability that such an electron leaves electrode 2 in a time interval of duration dt is then

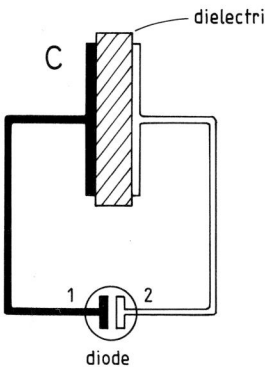

Figure 12.1 Alkemade's diode consisting of two electrodes of different materials with work functions W_1 and W_2. The diode is connected to a capacitor, as indicated. The two capacitor plates are made of the same materials as the two electrodes. The capacitance of the diode itself is considered negligible.

$$P_{II}\, dt = \frac{4\pi m}{h^3}(\kappa T)^2 \mathcal{A} e^{-W_2/(\kappa T)} e^{-e^2(N+1/2)/(\kappa TC)}\, dt$$
$$\equiv \xi e^{-(W_2-W_1)/(\kappa T)} e^{-e^2(N+1/2)/(\kappa TC)}\, dt. \qquad (12.3)$$

Note that $P_{II} = P_{II}(N)$, in contrast to P_I, which is independent of N. Let $P(N, t)$ denote the probability of there being N excess electrons on electrode 1 at a time t, where N can be negative in case of an electron deficit. At a time $t + \Delta t$ the probability can be $P(N)$ provided that

(i)   it was $P(N)$ at t and no electrons were emitted from either electrode,

(ii) it was $P(N-1)$ at t and one electron was emitted from electrode 2 while none was emitted from electrode 1,

(iii) it was $P(N+1)$ at t and one electron was emitted from electrode 1 while none was emitted from electrode 2, and

(iv) it was $P(N)$ at t and an electron was emitted from electrode 2 as well as one from electrode 1.

The probability of two or more electrons being emitted in Δt is considered negligible. It is assumed that emissions of electrons from the two electrodes are statistically independent. Consequently we have

$$P(N, t + \Delta t) = P(N, t)(1 - P_I\Delta t)[1 - P_{II}(N)\,\Delta t]$$
$$+ P(N-1, t)P_{II}(N-1)\,\Delta t\,(1 - P_I\,\Delta t)$$
$$+ P(N+1, t)P_I\,\Delta t\,[1 - P_{II}(N+1)\,\Delta t]$$
$$+ P(N, t)P_I\,\Delta t\,P_{II}(N)\,\Delta t,$$

in a self-evident notation. Ignoring terms containing $(\Delta t)^2$, with the approximation

$$\frac{\partial P(N, t)}{\partial t} \approx \frac{P(N, t + \Delta t) - P(N, t)}{\Delta t},$$

and by using the explicit expressions for P_I and P_{II}, the following master equation is obtained:

$$\frac{1}{\xi}\frac{\partial P(N, t)}{\partial t} = P(N+1, t) - P(N, t)$$
$$+ \zeta[e^{-(N-1)\varepsilon}P(N-1, t) - e^{-N\varepsilon}P(N, t)], \tag{12.4}$$

with the abbreviations

$$\varepsilon = \frac{e^2/C}{\kappa T}, \qquad \eta = \frac{W_1 - W_2}{\kappa T}, \qquad \zeta = e^{\eta - \varepsilon/2}, \tag{12.5}$$

while ξ was defined in (12.1). An equilibrium solution, $P_{eq}(N)$, is readily obtained by putting

$$P_{eq}(N+1) - \zeta e^{-N\varepsilon}P_{eq}(N) = P_{eq}(N) - \zeta e^{-(N-1)\varepsilon}P_{eq}(N-1), \tag{12.6}$$

which shows that $P_{eq}(N+1) - \zeta e^{-n\varepsilon}P_{eq}(N)$ must equal a constant independent of N, in particular also for $N \to \infty$, for which $P_{eq}(N) \to 0$ so that the constant must be zero. Consequently, by iteration

$$P_{eq}(N) = e^{-\varepsilon N^2/2 + \eta N}P_{eq}(0), \tag{12.7}$$

where $P_{eq}(0)$ is determined by the normalization condition

$$P_{eq}^{-1}(0) = \sum_{-\infty}^{\infty} e^{-\varepsilon N^2/2 + \eta N} \approx \sqrt{\frac{2\pi}{\varepsilon}} e^{\eta^2/(2\varepsilon)}.$$

Within the latter approximation (obtained by replacing the summation by an integral) it can readily be demonstrated that $P_{eq}(N)$ has its maximum at $\langle N \rangle_{eq} = \eta/\varepsilon = (W_1 - W_2)C/e^2$ and with mean-square fluctuations $\langle (N - \langle N \rangle_{eq})^2 \rangle_{eq} = 1/\varepsilon$. The mean-square value of the voltage fluctuations is $\kappa T/C$, in agreement with previous results.

- **Exercise:** Derive an equation for the moment-generating function $M(\alpha) = \sum_N P(N, t)\exp(-N\alpha)$ from (12.4).

It is now interesting to note that the nonlinearity of the diode indeed builds up a net charge, and thus a net voltage $-(e/C)\langle N \rangle_{eq} = -(W_1 - W_2)/e$, across the capacitor. This is the contact potential caused by the difference between the work functions. This voltage is not violating basic thermodynamics because the two halves of the circuit consist of materials with different work functions, and the capacitor can not be discharged to produce a net current driving an electromotor. A full treatment of the problem, with capacitor plates of the *same* materials, with differences in work function only in the diode itself, requires that one take into account also the contact potentials of the different metals in the connecting wires. This problem will not be discussed here.

A dynamic equation for the average number of charges, $\langle N \rangle$, is readily obtained by considering $d\langle N \rangle/dt \equiv d\sum NP(N)/dt$:

$$\frac{1}{\xi}\frac{d\langle N \rangle}{dt} = -1 + e^{\eta - \varepsilon/2}\langle e^{-\varepsilon N} \rangle. \tag{12.8}$$

Equations for higher order moments can be obtained similarly. Evidently, the equation for $\langle N \rangle$ can not be solved directly without knowledge of the average $\langle e^{-\varepsilon N} \rangle$. This will depend on the entire distribution $P(N)$ and also on C. However, the equation can be used to derive a relation for the macroscopic features of the diode in the limit of vanishing fluctuations $\varepsilon \to 0$ obtained formally by letting $C \to \infty$ and $N \to \infty$ in such a way that the voltage $-eN/C$ is fixed. On introducing the current $I = e\, d\langle N\rangle/dt$, the result is obtained in the form of the standard diode characteristic

$$I = e\xi(e^{eV/(\kappa T)} - 1), \tag{12.9}$$

where $V = W_1 - W_2 - eN/C$. (This is *not* the voltage across the capacitor.) The relative fluctuations around the average vanish as $N \to \infty$ and $C \to \infty$ so it is no longer necessary to distinguish between N and $\langle N\rangle$ in this limit.

12.2   The Fokker–Planck approximation

For physically realistic examples, the number of excess electrons on a capacitor plate can be taken to be so large that N can be considered a continuous rather than a discrete variable. In this limit $P(N)$ is a coarse-grained distribution in the sense discussed already in Chapter 8. The master equation (12.4) can be approximated by a Fokker–Planck equation. With the abbreviations introduced before, we have the probability of a decrease by one in the number N of electrons

$$P_{\mathrm{I}}\, dt = \xi\, dt,$$

and for an increase in number by one electron

$$P_{\mathrm{II}}dt = \xi e^{-\varepsilon(N+1/2)} e^{\eta} dt.$$

The coefficients in the Fokker–Planck equation are then calculated as

$$a_1 = -\xi + \xi e^{\eta} e^{-\varepsilon(N+1/2)},$$

$$a_2 = \xi + \xi e^{\eta} e^{-\varepsilon(N+1/2)},$$

as discussed before. The resulting Fokker–Planck equation becomes

$$\frac{1}{\xi}\frac{\partial P(N, t)}{\partial t} = \frac{1}{2}\frac{\partial^2}{\partial N^2}\left(1 + e^{\eta} e^{-\varepsilon(N+1/2)}\right) P(N, t)$$
$$- \frac{\partial}{\partial N}\left(-1 + e^{\eta} e^{-\varepsilon(N+1/2)}\right) P(N, t). \tag{12.10}$$

This equation has a zeroth-order or equilibrium solution

$$P_{\mathrm{FP}}(N) = P_0 e^{2(\eta - \varepsilon N)}(1 + e^{\eta} e^{-\varepsilon(N+1/2)})^{-1-4/\varepsilon}. \tag{12.11}$$

The solution is labeled FP to distinguish it from the solution (12.7) to the master equation (12.4). The constant P_0 is determined by normalizing $P_{\mathrm{FP}}(N)$. The result is noticeably different than (12.7) when it comes to details, but reproduces the lowest order moments to within an accuracy of $\mathcal{O}(\varepsilon)$ or better (van Kampen, 1960).

It is now convenient to introduce a substitution $N = \mathcal{N} + N'$, which can be considered as a shift of the zero point on the scale for N. On choosing $\mathcal{N} = \eta/\varepsilon$, the equation becomes

$$\frac{1}{\xi} \frac{\partial P(N', t)}{\partial t} = \frac{1}{2} \frac{\partial^2}{\partial N'^2} \left(1 + e^{-\varepsilon(N'+1/2)}\right) P(N', t)$$

$$- \frac{\partial}{\partial N'} \left(-1 + e^{-\varepsilon(N'+1/2)}\right) P(N', t). \tag{12.12}$$

To calculate the fluctuation spectrum, the equation for the first moment is obtained as

$$\frac{1}{\xi} \frac{d\langle N' \rangle}{dt} = -1 + e^{-\varepsilon/2} \langle e^{-\varepsilon N'} \rangle, \tag{12.13}$$

which could be obtained from (12.8) as well. The relation (12.13) requires an expression for the average $\langle e^{-\varepsilon N'} \rangle$. To obtain this in terms of its series expansion, equations for the higher moments are required. Also these can be obtained from (12.12) as

$$\frac{1}{\xi} \frac{d\langle N'^2 \rangle}{dt} = 1 + e^{-\varepsilon/2} \langle e^{-\varepsilon N'} \rangle - 2\langle N' \rangle + 2e^{-\varepsilon/2} \langle N' e^{-\varepsilon N'} \rangle, \tag{12.14}$$

$$\frac{1}{\xi} \frac{d\langle N'^3 \rangle}{dt} = -1 + e^{-\varepsilon N'/2} \langle e^{-\varepsilon N'} 3\langle N' \rangle + 3e^{-\varepsilon/2} \langle N' e^{-\varepsilon N'} \rangle - 3\langle N'^2 \rangle + 3e^{-\varepsilon/2} \langle N'^2 e^{-\varepsilon N'} \rangle, \tag{12.15}$$

etc. No exact analytic solution for these coupled equations can be found. Following van Kampen (1960), an approximate solution is sought in powers of ε. It is here an advantage to rescale the temporal variable by introducing a new time $\tau = \xi \varepsilon t$, implying that the timescale increases linearly with C, i.e. proportional to the R–C time constant of the circuit. In order to make the various powers of ε explicit, $\langle N'^p \rangle = \varepsilon^{-p/2} m_p$ is introduced, in expectation of $\langle N'^p \rangle = \mathcal{O}(\varepsilon^{-p/2})$ for $p = 1, 2, \ldots$ as suggested by the equilibrium solution (12.7), rescaled to the present variables:

$$\frac{dm_1}{d\tau} = -m_1 - \tfrac{1}{2} \varepsilon^{1/2} (1 - m_2) + \tfrac{1}{2} \varepsilon (m_1 - \tfrac{1}{3} m_3) + \mathcal{O}(\varepsilon^{3/2}),$$

$$\frac{dm_2}{d\tau} = 2 - 2m_2 - \varepsilon^{1/2} (2m_1 - m_3) + \mathcal{O}(\varepsilon),$$

$$\frac{dm_3}{d\tau} = -3m_3 + 6m_1 + \mathcal{O}(\varepsilon^{1/2}).$$

These equations are complete for $m_1(\tau)$, $m_2(\tau)$, and $m_3(\tau)$ up to order $\varepsilon^{3/2}$ and can be solved once the initial values $m_1(0)$, $m_2(0)$, and $m_3(0)$ are known. The results for m_j, with $j = 1, 2$, and 3, are *conditional*, in terms of these initial conditions. To zeroth order in ε the results are

$$m_1(\tau) = m_1(0) e^{-\tau},$$

$$m_3(\tau) = m_3(0) e^{-3\tau} + 3m_1(0)(e^{-\tau} - e^{-3\tau}).$$

To order $\varepsilon^{1/2}$ one finds

$$m_2(\tau) = m_2(0) e^{-2\tau} + 1 - e^{-2\tau} + \varepsilon^{1/2} [m_3(0) - 3m_1(0)](e^{-2\tau} - e^{-3\tau}) + \varepsilon^{1/2} m_1(0)(e^{-\tau} - e^{-2\tau}).$$

Finally, to order ε

$$m_1(\tau) = m_1(0) e^{-\tau} + \tfrac{1}{2} \varepsilon^{1/2} [m_2(0) - 1](e^{-\tau} - e^{-2\tau})$$

$$+ \varepsilon m_1(0)(\tfrac{1}{2} \tau e^{-\tau} - e^{-\tau} + 2e^{-2\tau} - e^{-3\tau})$$

$$+ \varepsilon m_3(0)(\tfrac{1}{6} e^{-\tau} - \tfrac{1}{2} e^{-2\tau} + \tfrac{1}{3} e^{-3\tau}), \tag{12.16}$$

where $m_1(0) = \varepsilon^{1/2} N_0$, $m_2(0) = \varepsilon N_0^2$, and $m_3(0) = \varepsilon^{3/2} N_0{}^3$ in terms of the initial value N_0. The correlation function can now be obtained by multiplying (12.16) by N_0 and averaging by use of the equilibrium probability density (12.11), with the shift of zero point on the scale for N used here. The averaging will require the calculation of only three terms, i.e. $\langle N_0 m_1(0) \rangle = \varepsilon^{1/2} \langle N_0^2 \rangle = 1/\varepsilon^{1/2} + \mathcal{O}(\varepsilon^{3/2})$ and $\langle N_0 m_3(0) \rangle = \varepsilon^{3/2} \langle N_0^4 \rangle = 3/\varepsilon^{1/2} + \mathcal{O}(\varepsilon^{1/2})$, while the term $\langle N_0 m_2(0) \rangle = \varepsilon \langle N_0^3 \rangle = \mathcal{O}(\varepsilon)$ will not contribute, to this order. The final result is

$$\varepsilon \langle N_0 N(\tau) \rangle = (1 - \tfrac{1}{2}\varepsilon)e^{-\tau} + \tfrac{1}{2}\varepsilon e^{-2\tau} + \tfrac{1}{2}\varepsilon\tau e^{-\tau}. \tag{12.17}$$

By Fourier transformation the power spectrum for the fluctuations is obtained as

$$\varepsilon S(\omega) = \frac{2}{\pi}\left(\frac{1-\varepsilon}{1+\omega^2} + \frac{\varepsilon}{4+\omega^2} + \frac{\varepsilon}{(1+\omega^2)^2}\right). \tag{12.18}$$

This spectrum is correct up to $\mathcal{O}(\varepsilon)$. The high-frequency limit of (12.18) gives a power spectrum proportional to ω^{-2}. For smaller frequencies, the spectrum is different than the one obtained for a Gaussian Markov process; the statistics of the present problem are non-Gaussian. The result here differs from that obtained by use of the fluctuation–dissipation theorem, as expected because that theorem is valid in the linear limit only, as has already been mentioned. The Wiener–Khinchine theorem used to obtain the spectrum on the basis of the correlation function has nothing to do with this question; it refers merely to a matter of representation.

- **Exercise:** Rewrite the power spectrum obtained in (12.18) in terms of the physical quantities.

- **Exercise:** Why had the N-axis to be rescaled? Could the solution in terms of the expansion ε have been obtained from (12.10) directly?

The fluctuation–dissipation theorem relates the auto-correlation function to the resistivity obtained from the slope of the diode characteristic (12.19) at $(I, V) = (0, 0)$. In the power expansion in ε, this operation corresponds to retaining the linear, first-order, equation

$$\frac{dm_1}{d\tau} = -m_1,$$

with the solution

$$m_1(\tau) = m_1(0)e^{-\tau}.$$

The auto-correlation function to this order in ε then becomes

$$\langle N_0 N(\tau) \rangle = \langle N_0^2 \rangle e^{-\tau} = \frac{1}{\varepsilon}e^{-\tau},$$

or

$$\varepsilon \langle N_0 N(\tau) \rangle = e^{-\tau} + \mathcal{O}(\varepsilon^2).$$

This is consistent with the general result for stationary *Gaussian* Markov processes, see (10.44), and the corresponding power spectrum, obtained by Fourier transformation according to the Wiener–Khinchine theorem, has the universal Lorentzian shape, see (10.45).

13   Fermi acceleration

In his studies of the origin of cosmic rays, Fermi proposed in 1949 that charged particles could be accelerated by reflections from propagating disturbances in space. In his discussions Fermi (1949) explicitly mentioned charged particles being reflected by moving magnetic mirrors (Chen, 1983), but the results are evidently much more general. Today one usually distinguishes between two types of mechanism of acceleration. The first is the so-called first-order acceleration whereby an electron or ion propagating with a velocity u is reflected by a head-on collision with, for instance, an electrostatic shock moving toward it with a speed V. For elastic reflections the particle is then reflected with a velocity of magnitude $u + 2V$. It is, however, not self-evident that particles are likely to encounter shocks with velocities, V, that are large enough to cause a substantial acceleration after just a single reflection, and Fermi also argued that there is a second-order acceleration process, whereby the energy gained by a particle is a consequence of many reflections, which individually give only a modest increase in energy. The latter mechanism will be discussed here.

It is trivially evident that particles can be accelerated by many small successive impacts if they each give a slight gain in energy. However, on reflection from a receding object, an overtaking collision, a *loss* of energy will be experienced. In terms of the foregoing example, the speed of the outgoing charged particle will be $u - 2V$. In the case of reflections from randomly moving objects, chocks, mirrors etc., one might naively expect that reflection from an approaching and a receding object should be equally probable, and therefore no *net* acceleration would be observed. This is not so, as the following example will illustrate. What matters is the relative velocities, and reflection from an approaching object is slightly more probable than reflection from a receding one. The model described in the following was first proposed by Hammersley (1961), but various aspects of the problem have been elaborated also by many other authors.

13.1   The stochastic model

The model considered here is essentially one-dimensional and can be described in terms of a simple time–position diagram such as the one shown in Fig. 13.1. A particle moving with constant velocity u appears in this diagram as a straight line with slope u. Figure 13.1 shows the trajectory of a massive particle, or rather a piston, moving with a sawtooth-like oscillatory motion with a constant velocity of magnitude either $+\frac{1}{2}$ or $-\frac{1}{2}$. The period of oscillation is taken to be t_0 and the peak-to-peak amplitude of the oscillation is then $t_0/2$. Assume now that a light particle is colliding with the heavy one. Let the velocity of the light particle be u before impact. After the collision, its velocity will be $u + 1$ if it happens to be reflected head-on, or

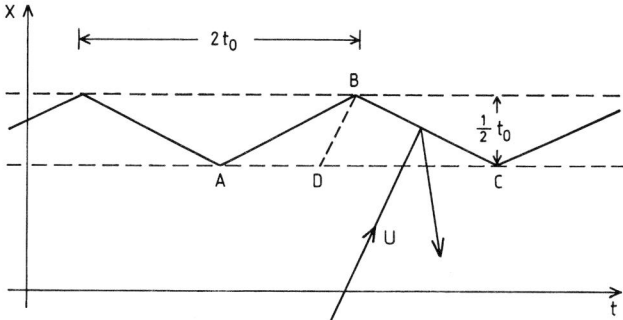

Figure 13.1 A space–time diagram for illustration of the second-order Fermi-acceleration process.

alternatively the velocity is $u - 1$ in case of an overtaking collision. In order to calculate the respective probabilities of the two types of reflection, it is now assumed that the light particle arrives at the position of the heavy one with equal probability within any time interval $t, t + \delta t$. By inspection of Fig. 13.1 it is readily seen that the transition probabilities $u \to u + 1$ and $u \to u - 1$ are in the ratio $DC = t_0[1 + 1/(2u)]$ to $AD = t_0[1 - 1/(2u)]$, with the definitions of Fig. 13.1, provided that $u > \frac{1}{2}$. Given that a reflection occurs, the conditional probabilities are then

$$P(u \to u + 1) = \frac{1}{2} + \frac{1}{4u},$$
$$P(u \to u - 1) = \frac{1}{2} - \frac{1}{4u}, \tag{13.1}$$

for $u > \frac{1}{2}$. It will be demonstrated that, for properly chosen initial conditions, this inequality is always satisfied.

The probability of a collision within a time interval is assumed to follow a Poisson distribution, i.e. the probability of a collision within a small time interval, dt, is proportional to dt itself with a coefficient equal to the velocity u multiplied by a constant. With a proper choice of time unit, the constant of proportionality can be taken to be 2. Physically, the proportionality to u is intuitively clear, i.e. a fast particle 'covers more ground' than does a slow one and has a larger probability of encountering a reflector, the heavy particle in Fig. 13.1. Assuming that reflectors are uniformly distributed in space, and with constant geometric cross-sections, the proportionality to v then follows automatically. The probability of a collision within a time interval of duration dt is thus $2u \, dt$. Consequently, using (13.1), the probabilities relating a velocity u at time t to the velocity at $t + dt$ are

$$P(u + 1|u) = \tfrac{1}{2}(2u + 1) \, dt,$$
$$P(u|u) = 1 - 2u \, dt,$$
$$P(u - 1|u) = \tfrac{1}{2}(2u - 1) \, dt. \tag{13.2}$$

These are the transition probabilities for a Markov process. Note that the same results can be argued by assuming the probability of a collision to be proportional to the relative velocity, here $u \pm \frac{1}{2}$. To simplify the problem it is now assumed that the light particle starts out with the velocity $\frac{1}{2}$. All later velocities can then be written as $\frac{1}{2} + k$ with k an integer. With a little algebra, the probability of the particle being in the state k is obtained as

$$P_k(t + dt) = [1 - (2k + 1)\, dt]P_k(t) + (k\, dt)P_{k-1}(t) + [(k + 1)\, dt]P_{k+1}(t). \tag{13.3}$$

It is evident that, with the present assumptions, we have $u \geq \frac{1}{2}$ at all times, as was assumed from the outset. For infinitesimally small dt the relation (13.3) is reduced to the differential equation which is equivalent to a master equation

$$\frac{dP_k(t)}{dt} = -(2k + 1)P_k(t) + kP_{k-1}(t) + (k + 1)P_{k+1}(t), \tag{13.4}$$

for $k = 0, 1, 2, \ldots$. The initial condition is $P_0(0) = 1$ and $P_{k>0}(0) = 0$. Evidently, to determine the evolution of $P_k(t)$ we have to determine the temporal variation of $P_{k-1}(t)$, which depends on $P_{k-2}(t)$ etc. It turns out (Hammersley, 1961) that the complication associated with this hierarchy of equations can be circumvented by reformulating (13.4) in terms of the generating function,

$$G(x) \equiv \langle x^k \rangle = \sum_0^\infty P_k(t)x^k. \tag{13.5}$$

This function has properties similar to those of the moment-generating function. After simple manipulations one obtains from (13.4) the relation

$$\frac{\partial G(x, t)}{\partial t} = (1 - x)^2\, \frac{\partial G(x, t)}{\partial x} - (1 - x)G(x, t), \tag{13.6}$$

for $0 \leq x < \infty$. The equation (13.6) has to be solved with the self-evident initial condition $G(x, 0) = 1$. The solution can be obtained as

$$G(x, t) = \frac{1}{1 + (1 - x)t}, \tag{13.7}$$

as is demonstrated by insertion into (13.6).

- **Exercise:** Obtain the general solution of (13.6) for arbitrary initial conditions.

- **Exercise:** From (13.4), derive an equation for the moment-generating function $M(\alpha, t) = \sum_k P_k(t)\exp(\alpha k)$ defined in Chapter 2.

It is readily demonstrated by use of the definition of G in (13.5) that the nth factorial moment

$$\left.\frac{\partial^n G(x, t)}{\partial x^n}\right|_{x=1} = \langle (k - n + 1) \cdots (k - 1) \cdot k \rangle \equiv \left\langle \frac{k!}{(k - n)!} \right\rangle.$$

On the other hand, we can differentiate the result (13.7) n times to obtain

$$\left.\frac{\partial^n G(x, t)}{\partial x^n}\right|_{x=1} = n!\, t^n,$$

implying that

$$\left\langle \frac{k!}{(k - n)!} \right\rangle = n!\, t^n. \tag{13.8}$$

In particular, the average state number is $\langle k \rangle = t$. The variance of state numbers is

$$\langle k^2 \rangle - \langle k \rangle^2 = \left\langle \frac{k!}{(k - 2)!} \right\rangle + \left\langle \frac{k!}{(k - 1)!} \right\rangle - \left\langle \frac{k!}{(k - 1)!} \right\rangle^2 = t(t + 1),$$

since, for instance, $\langle k!/(k-1)!\rangle = \langle k!k/[(k-1)!k]\rangle = \langle k\rangle$, etc.

The light particle, or a system of many independent light particles, will thus never reach equilibrium, but have a steadily increasing velocity, on average, and also an increasing scatter in the occupied states. This is after all to be expected; in the present model the heavy particle receives momentum but not energy when it is reflecting a light particle. Its mass being infinite, the heavy particle has infinite kinetic energy and represents an unlimited source of energy. Equilibration will be effective for realistic systems when the average kinetic energy, $\frac{1}{2}m\langle u^2\rangle$, of the light particle approaches a finite kinetic energy, $\frac{1}{2}MV^2$, of the heavy particle.

To illustrate these features in more detail, we can obtain the analytic result for $P_k(t)$ by making a series expansion of (13.7) around $x = 1$ and find

$$G(x, t) = \frac{1}{1+t} \sum_n \frac{(xt)^n}{(1+t)^n}.$$

(13.9)

By identification of this term by term, using the definition (13.5), we find

$$P_k(t) = \frac{t^k}{(1+t)^{k+1}}.$$

(13.10)

Consequently

$$\frac{P_{k+1}(t)}{P_k(t)} = \frac{t}{(1+t)} < 1,$$

(13.11)

and the probability at any given time of the occupation of any given state is a decreasing function of the state number. The most likely state to be occupied is therefore the ground state, corresponding to $k = 0$. For a process of duration T, the expected time spent in a state k is obtained as $T_k = \int_0^T P_k(t)\,dt$, which can be evaluated after some algebra using (13.10) to be

$$T_k = \log(1 + T) - \sum_{r=1}^{k} \frac{T^r}{r(1+T)^r},$$

meaning that

$$\log(1 + T) - \sum_{r=1}^{k} \frac{1}{r} \le T_k \le \log(1 + T).$$

Hence, for any fixed k, we have $T_k \sim \log T$ as $T \to \infty$. From the second relation in (13.2) it is concluded that $P(k|k) = 1 - (2k + 1)\,dt$ since $u = \frac{1}{2} + k$. If the light particle has just arrived at a state labeled k, the expected time it will remain there is then $1/(2k + 1)$. Since this time is finite, whereas T_k tends to infinity with time $\to \infty$, it is concluded that the state with label k is visited infinitely often, when the process continues indefinitely (Hammersley, 1961).

It might be instructive briefly to consider the situation in which k is considered as a continuous variable, i.e. in a coarse-grained description of the probability density. The following equation is readily obtained:

$$\frac{\partial P(k, t)}{\partial t} = k\frac{\partial^2 P(k, t)}{\partial k^2} + \frac{\partial P(k, t)}{\partial k},$$

or

$$\frac{\partial P(k, t)}{\partial t} = \frac{\partial}{\partial k}k\frac{\partial P(k, t)}{\partial k}.$$

(13.12)

This equation has the steady-state solutions $P(k) =$ constant or $P(k) = \beta \ln k$, where β is a constant. Neither of these solutions can be normalized, so we conclude that the problem in this formulation does not have any steady-state solution either. It is straightforward to obtain $\langle k \rangle = t$, which agrees with the result from (13.7). From (13.12) we also find $\langle k^2 \rangle - \langle k \rangle^2 = t^2$, which agrees with the result from (13.7) for large times.

- **Exercise:** Derive a differential equation for $P(u, t)$ for the case in which the velocity V of the reflector is varying continuously with a known distribution $P(V)$. Assume that V is slowly varying so that it can be taken constant during the duration of the reflection.

The foregoing results have rather interesting implications. The heavy particle can be considered as a moving piston in a cylinder containing a gas of light particles. The system is isolated from the surroundings and the processes in the gas are adiabatic, or isentropic. From classical thermodynamics we know that, after a full cycle, the system is back in its original state for adiabatic processes, i.e. in half of the cycle, during the expansion phase, the gas does work on the piston. This energy is given back during the second phase, the compression. Now, the system illustrated in Fig. 13.1 is in principle equivalent to the piston and it undergoes many cycles, each of duration t_0. However, it is evident, and the conclusion seemingly inescapable by simple physical reasoning, that the net energy of the gas is steadily increasing, in apparent contradiction to the thermodynamic result just quoted. In reality, however, an important restriction on the validity of the thermodynamic analysis was omitted; classical thermodynamics operates with states that are always in local equilibrium, or, in other terms, with pistons moving very slowly, i.e. $t_0 \to \infty$. The results summarized in this section violate this condition and can be seen as describing 'finite-time' thermodynamics.

13.2   Deterministic models

The model in the foregoing section contained the basic assumption that the light particle arrived at the moving piston with a random uniformly distributed phase. This is in many ways a reasonable assumption and many scenarios having such properties can be envisaged; a sort of pin-ball game, for instance. A simple closed model will, on the other hand, not necessarily have this feature, and it is interesting to analyze also such a deterministic model. Two candidates are illustrated in Fig. 13.2. In one the light particle bounces back and forth between an oscillating wall and a fixed one separated by a distance ℓ. The particle's velocity is assumed constant during the transit from the oscillating wall to the fixed wall. In the other model, the light particle repeatedly returns to the oscillating wall due to the action of a constant gravitational acceleration g. With these assumptions and given specific initial conditions, we can calculate the velocity of the light particle after each encounter with the moving wall. The result can be understood as a mapping of the velocity w_n of the light particle onto a new velocity w_{n+1} at the nth reflection from the oscillating wall, or piston.

The two models can again be described adequately in one spatial dimension with the piston's displacement assumed to be a known function of time, for instance the sawtooth function assumed in Fig. 13.1. The light particle is most conveniently described in a phase

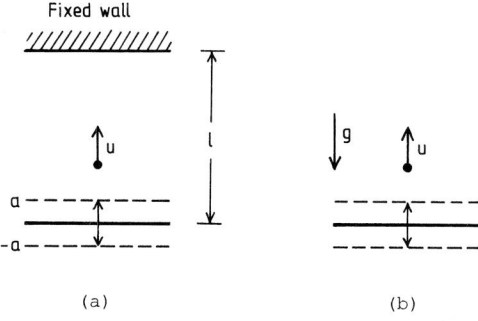

Fixed wall

(a)   (b)

Figure 13.2 An illustration of a deterministic model for Fermi acceleration. In (a) the light particle bounces back and forth between an oscillating wall and a fixed one separated by a distance ℓ. In model (b), the light particle repeatedly returns to the oscillating wall due to the action of a constant gravitational acceleration g. In model (a), the effect of gravity is usually ignored.

space spanned by its velocity and the phase with which it hits the moving wall. The analysis will in general become rather complicated. For instance, for the situation illustrated in Fig. 13.2(a), assuming that we have the sawtooth function as in Fig. 13.1, Zaslavskii and Chirikov (1964) obtained the exact mapping as

$$u_{n+1} = \pm u_n + \psi_n - \tfrac{1}{2}, \tag{13.13}$$

$$\psi_{n+1} = \tfrac{1}{2} - 2u_{n+1} + \left[\left(\tfrac{1}{2} - 2u_{n+1} \right)^2 + 4\phi_n u_{n+1} \right]^{1/2}, \quad \text{for } u_{n+1} > \tfrac{1}{4}\psi_n, \tag{13.14}$$

$$\psi_{n+1} = 1 - \psi_n + 4u_{n+1}, \qquad \text{for } u_{n+1} \leq \tfrac{1}{4}\psi_n, \tag{13.15}$$

$$\phi_n = \left\{ \psi_n + \frac{\psi_n(1 - \psi_n) + \ell/(4a)}{4u_{n+1}} \right\}, \tag{13.16}$$

where $2a$ is the peak amplitude of the piston's displacement, ℓ is the minimum distance between the wall and the piston, $\tfrac{1}{4}V$ is the absolute value of the wall's velocity, and ψ is the phase of the piston, changing from 0 to $\tfrac{1}{2}$ in the first half cycle and from $\tfrac{1}{2}$ to 1 in the reverse motion. The particle's velocity u is normalized with V and n is the number of the actual collision with the oscillating piston. The brackets { } in (13.16) denote the fractional part of the argument. It is quite evident that a simplification would be appreciated. This can be obtained by allowing the oscillating piston to impart momentum to the light particle according to the actual velocity of the piston, but ignoring its change in spatial position. With this approximation, the set of equations (13.13)–(13.16) is reduced to

$$u_{n+1} = \left| u_n + \psi_n - \tfrac{1}{2} \right|,$$
$$\psi_{n+1} = \left\{ \psi_n + \frac{K}{u_{n+1}} \right\}, \tag{13.17}$$

where $K = \ell/(16a)$, $K/u = 2\ell/(wT)$ is the normalized transit time, $T = 32a/V$ is the oscillation period of the piston, and $w = uV$ is the particle's velocity. An absolute-value sign was included in a rather *ad hoc* manner to correspond to the reversal of velocity at low velocities, $u < 1$, which appears in the exact mapping (13.13)–(13.16).

The analysis can readily be generalized also to other temporal variations of the piston's displacement. It can for instance be taken to be sinusoidal, with the resulting mapping

$$u_{n+1} = |u_n + \sin \psi_n|,$$
$$\psi_{n+1} = \psi_n + \frac{2\pi K}{u_{n+1}}, \tag{13.18}$$

with the phase extending over 2π rather than unity. The form (13.18) is often called the *Ulam mapping* (Ulam, 1961).

We now consider the alternative situation illustrated in Fig. 13.2b. With a sinusoidal temporal variation of the piston's displacement and the same approximation as in (13.18), we find

$$u_{n+1} = u_n + \sin \psi_n,$$
$$\psi_{n+1} = \psi_n + K u_{n+1}, \tag{13.19}$$

but here with $K = 4\omega^2 a/g$, where ω is the frequency of the oscillating motion of the piston. The form (13.19) is often called the *standard map* or the *Pustylnikov mapping* (Pustylnikov, 1978). The two mappings (13.18) and (13.19) describe the basic difference in the physics of the problem; in the first case the time taken for transit through the system decreases as the light particle is accelerated, whereas we have the reverse situation in the other case. The Ulam and Pustylnikov mappings as well as other similar maps can be expressed in terms of the more general *radial twist mapping*

$$u_{n+1} = u_n + \sin \psi_n,$$
$$\psi_{n+1} = \psi_n + G(u_{n+1}), \tag{13.20}$$

with proper choice of the function G. These and similar maps have been studied extensively by, for instance, Lieberman and Lichtenberg (1972) and Lichtenberg *et al.* (1980).

- **Exercise:** Derive the Ulam and standard maps directly by considering Fig. 13.2.

- **Exercise:** Demonstrate that the standard map can be a local approximation of the Ulam map.

- **Exercise:** Determine the fixed points for the standard map and investigate their stability.

Typical results for the two different mappings are shown in Figs. 13.3 and 13.4. The conspicuous feature is the chaotic behavior, which is particularly apparent in Fig. 13.3 for small values of u. At first sight it might appear surprising that such relatively simple-looking mappings as (13.18) and (13.19) can have such a complicated behavior. In these regions the evolution of the velocity as a function of the number of reflection n is, however, extremely sensitive to the initial conditions. Even a minute difference in the initial value u_0 results in a quite different series of u_n's, see for instance Fig. 13.5, showing (u_n, ψ_n) for 15 reflections obtained from the Pustylnikov mapping using $K = 5$, with two slightly different initial conditions, i.e. $(u_0, \psi_0) = (0.5, 1)$ for the asterisks and $(u_0, \psi_0) = (0.500\,001, 1)$ for the filled circles. Note how the two solutions follow one another closely up to point number 11, then begin to separate slightly, and, at point number 14 the phase difference is close to 2π, although u is close to 6 for both conditions. Eventually, at point number 15, the separation in phase exceeds 2π and also the separation in u begins to increase. The results of the chaotic mappings are also

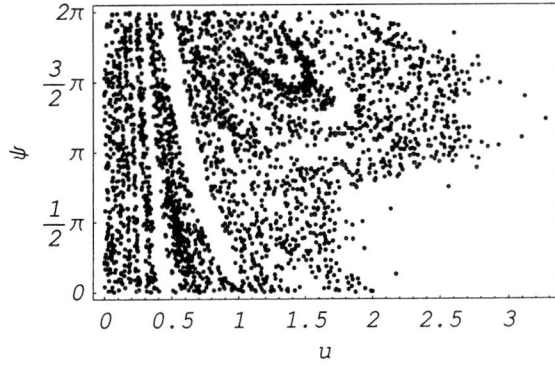

Figure 13.3 A diagram showing (u_n, ψ_n) for 3000 reflections obtained from the Ulam mapping with $2\pi K = 5$. The starting point is $(u_0, \psi_0) = (0.75, 1.25)$.

sensitive to the accuracy of the calculations; a calculation with 16-digit accuracy gives rise to the same visual appearance as a calculation with 12-digit accuracy, but without a point-to-point correspondence. Physically, this chaotic behavior arises because in the Ulam mapping it takes a long time, $\tau \approx 2\ell/w$, before the light particle encounters the moving piston. If $\tau \gg 1/\omega$ the phases of the piston at the nth and $(n + 1)$th encounters will in general be very different and the kicks the particle receives will appear as if they were statistically independent.

In the stochastic region the temporal variation of the velocity follows a Fokker–Planck equation to high accuracy in a coarse-grained description (Lichtenberg et al., 1980). There are islands of regular motion imbedded in the chaotic sea, but, if their widths are small, they will have only an insignificant influence on the diffusion in phase space. As the light particle's velocity is increased, its encounter with the piston happens at closely spaced intervals in time such that the phase of the piston's motion changes only moderately, and the motion (for the Ulam problem) becomes regular.

The Ulam mapping was analyzed by a combination of analytic and numerical work by Zaslavskii and Chirikov (1964), Lieberman and Lichtenberg (1972), and others. They showed that phase space as defined also in this chapter can be divided into three regions: (i) a stochastic

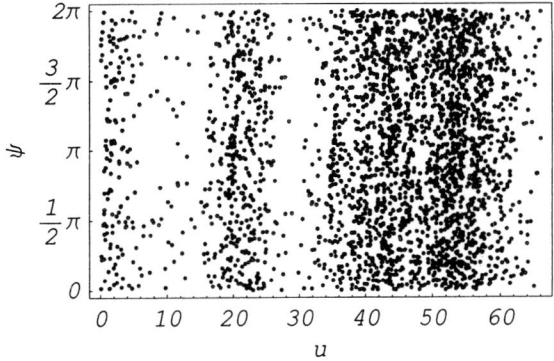

Figure 13.4 A diagram showing (u_n, ψ_n) for 3000 reflections obtained from the Pustylnikov mapping with $K = 5$. The starting point is $(u_0, \psi_0) = (2.5, 1.0)$.

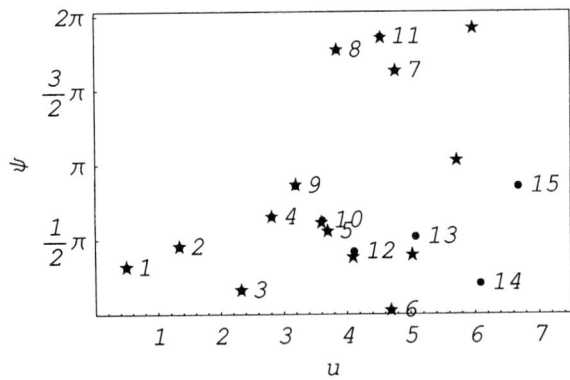

Figure 13.5 A diagram showing (u_n, ψ_n) for 15 reflections obtained from the Pustylnikov mapping using $K = 5$, with two slightly different initial conditions, i.e. $(u_0, \psi_0) = (0.5, 1)$ for the asterisks and $(u_0, \psi_0) = (0.500\,001, 1)$ for the filled circles.

region, essentially for low velocities; (ii) an intermediate-velocity range, within which islands of regular motion are imbedded in a stochastic sea; and (iii) a region at high velocities, for which most of the trajectories are regular and bounded. If the periodic piston velocity has a sufficient number of continuous derivatives, as in the present case with harmonic time variation (but not the sawtooth piston motion), invariant curves span the phase space, separating the stochastic sea of regions (i) and (ii) from region (iii), and thus bound the gain in energy of the light particle for that particular problem. This result is in agreement with the Kolmogorov–Arnold–Moser (KAM) theorem, which states that, for a dynamical system sufficiently close to integrable ones, most of the regular surfaces characteristic of integral systems continue to exist (Lichtenberg *et al.*, 1980). A KAM surface isolates higher velocities from stochastic motion at lower velocities.

For the Pustylnikov problem, on the other hand, an unbounded motion of the light particle can be observed. The existence of a KAM surface was indicated by Greene (1979) for $K < K_c \approx 0.97$ in (13.19). This implies that no globally isolating KAM surface exists for any velocity u, provided that

$$\frac{4\omega^2 a}{g} > 0.97. \tag{13.21}$$

In this case the light particle can be accelerated to unlimited velocity, very much like in the case analyzed by Hammersley (1961) which was discussed before.

The discussions in this chapter are relevant to many diverse problems. For instance Kendall (1961) has made an interesting application to the theory of comets, their escape from the solar system in particular.

Appendix A   The binomial distribution

Consider a discrete random variable that can assume a finite set of values, N. Assume an event having a probability s, with $0 < s < 1$. On performing N *independent* experiments there will be a certain probability P_r of the occurrence of r events, $0 \leq r \leq N$. To obtain an analytic expression for this probability, we argue as follows. The probability of exactly r occurrences of the desired event (successes) and $N - r$ failures in a particular order, out of N mutually independent trials, is given by $s^r q^{N-r} \equiv s^r (1-s)^{N-r}$ since the respective probabilities of success, s, and failure, $q = 1 - s$, are multiplied together when the events are independent of each other. Next, we must recognize that the number of different ways in which r successes can be divided among N trials is given by the binomial coefficient

$$\binom{N}{r} \equiv \frac{N(N-1)\cdots(N-r+1)}{1 \cdot 2 \cdots r} \equiv \frac{N!}{r!(N-r)!}$$

as the number of possible combinations of N trials taken r at a time. Thus, the probability of exactly r successes occurring among N trials is

$$P_r = s^r (1-s)^{N-r} \binom{N}{r}. \tag{A.1}$$

This is the *binomial distribution*. Note that $\sum_{r=0}^{N} P_r = 1$ as it should be. In Fig. A.1 we show P_r as a function of r for various combinations of s and N.

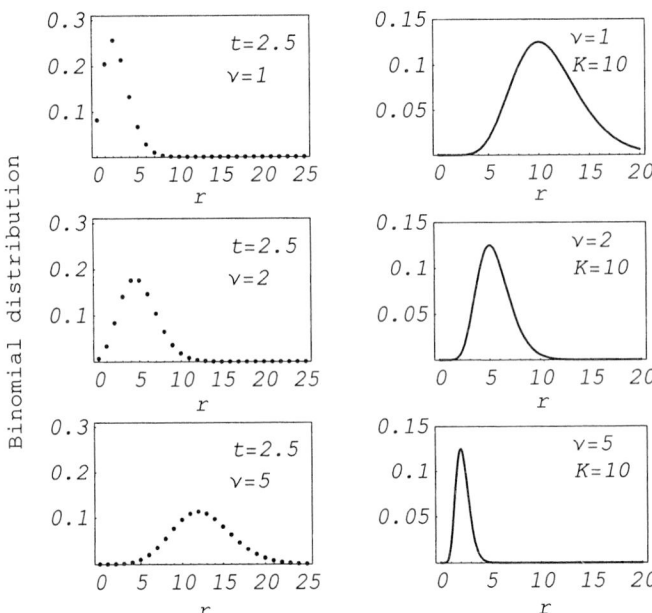

Figure A.1 The binomial distribution P_r as a function of r for various combinations of s and N.

Appendix B   The Poisson distribution

Consider a discrete random variable that can assume an infinite set of values. In this appendix we take the collisional model from Chapter 3 as an example and elaborate it in more detail. Using the same notation, we let the probability of an electron colliding with a neutral species within a short time interval $\{t; t + \Delta t\}$ be

$$P(\text{one collision in } \Delta t) = v\,\Delta t, \tag{B.1}$$

where v is a constant. The number of collisions within a time interval of *finite* duration can take any (integer) number, in principle. In order to present an alternative derivation of the results in Chapter 3, we let the time interval t be written as a sum of sub-intervals Δt, giving

$$
\begin{aligned}
P(\text{no collision in } \{t_1; t_1 + n\Delta t\}) &= P(\text{no collision in } \{t_1; t_1 + \Delta t\} \\
&\quad \text{nor in } \{t_1 + \Delta t; t_1 + 2\,\Delta t\} \\
&\quad \text{nor in } \{t_1 + 2\Delta t; t_1 + 3\,\Delta t\}\dots\text{etc.)} \\
&= (1 - v\,\Delta t)^n \\
&= (1 - vt/n)^n \to e^{-vt}, \qquad \text{for } n \to \infty,
\end{aligned}
$$

with events belonging to two nonoverlapping time intervals being statistically independent. The probability of the electron undergoing exactly K collisions in a time interval $t + \Delta t$ can be obtained by arguing that

$$
\begin{aligned}
P(K \text{ collisions in } t + \Delta t) =\ &P(K - 1 \text{ collisions in } t \\
&\text{and one collision in } \Delta t) \\
&+ P(K \text{ collisions in } t \\
&\text{and no collisions in } \Delta t).
\end{aligned}
$$

Referring to the discussion of Markov processes, we have that the collision in the time interval Δt is statistically independent of previous collisions and consequently

$$P(K, t + \Delta t) = P(K - 1, t)P(1, \Delta t) + P(K, t)P(0, \Delta t),$$

giving

$$\frac{P(K, t + \Delta t) - P(K, t)}{\Delta t} + vP(K, t) = vP(K - 1, t).$$

In the limit $\Delta t \to 0$ again a differential equation results:

$$\frac{dP(K, t)}{dt} + vP(K, t) = vP(K - 1, t),$$

with the solution

$$P(K, t) = ve^{-vt} \int_0^t e^{v\tau} P(K - 1, \tau)\, d\tau.$$

Since $P(K = 0, t) = e^{-vt}$ is known, the result for any K can be iterated to give the Poisson distribution

$$P(K, t) = \frac{(vt)^K e^{-vt}}{K!}. \tag{B.2}$$

Here, $P(K, t)$ is consequently the probability of an electron colliding exactly K times in a time interval of duration t. Using (B.2) it is readily shown that $\langle K \rangle = vt$ and $\langle K^2 \rangle = (vt)^2 + vt$, etc. These averages are *conditional* in the sense that t is given. Alternatively, the expression (B.2) can be interpreted as the probability density for time intervals with duration t and containing exactly K collisions. In Fig. B.1(a) we show $P(K, t)$ as a function of K for various values of the collision frequency v and fixed times t. In Fig. B.1(b) we have $P(K, t)$ as a function of t for fixed K for the same values of v. The probability (B.2) is normalized at all times, i.e. $\sum_K P(K, t) = 1$ but $\int_0^\infty P(K, t)\, dt = 1/v$.

- **Exercise:** Show that, for shot noise, the distribution of the time interval, t, between an event and its Kth successor is

$$w_K(t) = \frac{v^K t^{K-1}}{(K-1)!} e^{-vt}. \tag{B.3}$$

  The probability of the Kth event occurring in the interval $\{t, t + dt\}$ is consequently $w_K(t)\, dt$. Demonstrate that $\int_{-\infty}^\infty t w_K(t)\, dt = K/v$ and $\int_{-\infty}^\infty t^2 w_K(t)\, dt = K(K+1)/v^2$.

- **Example:** A particle undergoing a random walk is undergoing collisions at times t_n with $n = 0, \pm 1, \pm 2, \ldots$ as shown in Fig. B.2. In addition the origin of the time axis,

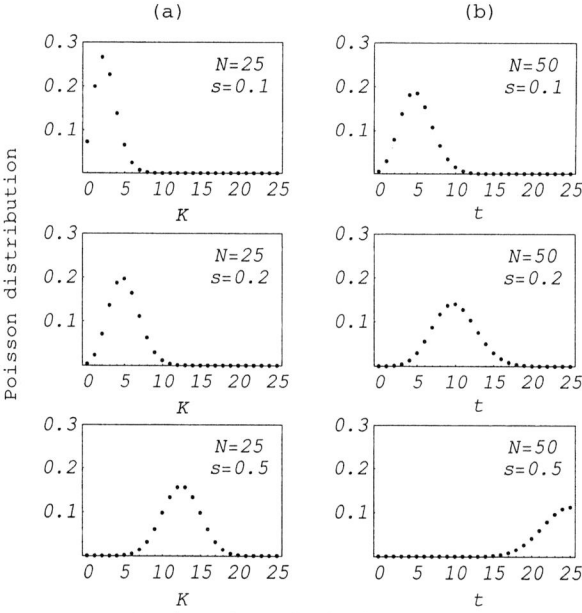

Figure B.1 (a) The Poisson distribution $P(K, t)$ as a function of the discrete variable K for $v = 1, 2$, and 5, with $t = 2.5$ fixed. (b) The Poisson distribution $P(K, t)$ as a function of the continuous variable t for fixed $K = 10$, for the same values of v.

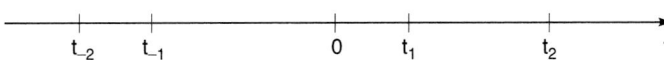

Figure B.2. The time axis, showing the collision times of a particle undergoing a random walk.

$\mathcal{O}$, is marked also, with $t_{-1} < \mathcal{O} < t_1$. The process is described by a Poisson distribution just like in the analysis in Chapter 3. The average $\langle t_n - t_{n-1} \rangle = 1/v$ as obtained before. On the other hand, using the basic definition of the probability of a collision in a time interval dt being $v\, dt$, we also find $\langle t_1 \rangle = 1/v$ and similarly $\langle t_{-1} \rangle = 1/v$, giving in general $\langle t_n - t_{n-1} \rangle = 2/v$, in contradiction to the foregoing standard result. One might expect that, in the present case, the additivity of averages would not be satisfied. This is, however, not so, and the seemingly paradoxical result can easily be understood by a geometric interpretation of the positioning of the origin in Fig. B.2. Selecting or placing the origin on the time axis, or selecting any other point for that matter, is not an unbiased process. Selecting a special interval for placing the origin can be seen as a collision process; the probability of hitting a certain interval is proportional (i) to its 'cross-section' (i.e. its length), and (ii) to the relative number of occurrences of that particular interval length (i.e. to its probability density). The probability of selecting a time interval with duration $\{t; t + dt\}$ is then given by $CtP(t)\, dt$ with $P(t) = \exp(-vt)$. The constant C is determined by normalizing the probability density. Consequently we find the average length of the time intervals thus selected to be

$$\langle t \rangle = \frac{\int_0^\infty t^2 P(t)\, dt}{\int_0^\infty t P(t)\, dt} = \frac{2}{v}.$$

The probability of finding a long time interval when selecting a certain time to be the origin of the time axis is larger than that of finding a short one! If, on the other hand, the time interval can be selected without its length having any relevance, then the probability of selecting one with a length within $\{t; t + dt\}$ is given simply by $P(t)\, dt = \exp(-vt)\, dt$ and we have $\langle t \rangle = 1/v$ as before. A thorough analysis of the problem discussed here is given by e.g. Feller (1971).

The Poisson distribution can be considered as a limiting case of the binomial distribution when $N \to \infty$. This can be demonstrated by first setting $s = d/D \ll 1$ and considering what happens to P_r when N and D approach infinity in such a way that their ratio $\lambda = N/D$ remains constant. By insertion into the binomial distribution we obtain

$$P_r = \frac{N!}{r!(N-r)!N^r} \left(\frac{Nd}{D}\right)^r \left(1 - \frac{Ns}{N}\right)^{N-r}$$

$$= \frac{\left(1 - \frac{1}{N}\right)\left(1 - \frac{2}{N}\right)\cdots\left(1 - \frac{r-1}{N}\right)}{r!} (\lambda d)^r \left(1 - \frac{\lambda d}{N}\right)^{N-r}. \tag{B.4}$$

As $N \to \infty$ we obtain the Poisson distribution, using the limiting form

$$e^{-v} = \lim_{N \to \infty} (1 - v/N)^N.$$

Appendix C   The Gaussian or normal distribution

Let $Y_1, Y_2, Y_3, \ldots, Y_N$ be N random variables with nonzero mean values. By an N-dimensional Gaussian random process we understand one with the property that its probability density function for all N has an N-dimensional normal form

$$P_N(y_1; \ldots; y_N) = \frac{1}{(2\pi)^{N/2}\sqrt{|A|}} \exp\left(-\frac{1}{2|A|}\sum_{i,j=1}^{N} A_{ij}(y_i - m_i)(y_j - m_j)\right), \tag{C.1}$$

with $m_i = \langle Y_i \rangle$, while A_{ij} is the cofactor of the second moment $d_{ij} = \langle (Y_i - \langle Y_i \rangle)(Y_j - \langle Y_j \rangle)\rangle$ in the matrix

$$\|A\| = \begin{Vmatrix} d_{11} & d_{12} & \cdots & d_{1N} \\ d_{21} & d_{22} & \cdots & d_{2N} \\ \vdots & \vdots & \vdots & \vdots \\ d_{N1} & d_{N2} & & d_{NN} \end{Vmatrix}, \tag{C.2}$$

and $|A|$ is the determinant of the matrix. Evidently, we are here dealing with square matrices, i.e. ones with the same number of rows and columns. The cofactor A_{ij} of any element d_{ij} is defined to be the determinant of order $N - 1$ formed by omitting the ith row and jth column of $\|A\|$ multiplied by $(-1)^{i+j}$. An N-dimensional Gaussian probability density is completely described by its average value (which can be taken to be zero, by a proper change of variables) and its second moments d_{ij}.

For illustration we consider first $N = 1$,

$$P_1(y_1) = \frac{1}{\sqrt{2\pi|A|}} \exp\left(-\frac{A_{11}}{2|A|}y_1^2\right), \tag{C.3}$$

where, for this simple case, $\|A\| = \|d_{11}\|$, and $|A| = d_{11}$, while $A_{11} = 1$, implying that $P_1(y_1) = (2\pi d_{11})^{-1/2}\exp(-\frac{1}{2}y_1^2/d_{11})$. A less trivial case is $N = 2$, for which

$$P_2(y_1; y_2) = \frac{1}{2\pi\sqrt{|A|}} \exp\left(-\frac{1}{2|A|}(A_{11}y_1^2 + A_{22}y_2^2 + A_{12}y_1y_2 + A_{21}y_2y_1)\right). \tag{C.4}$$

Now the correlation matrix becomes

$$\|A\| = \begin{Vmatrix} d_{11} & d_{12} \\ d_{21} & d_{22} \end{Vmatrix}, \tag{C.5}$$

and $|A| = d_{11}d_{22} - d_{12}d_{21}$ is the determinant of the matrix, $A_{11} = d_{22}$, $A_{22} = d_{11}$, and $A_{12} = A_{21} = -d_{12} = -d_{21}$. The projections $\int_{-\infty}^{\infty} P(y_1; y_2)\, dy_2$ and $\int_{-\infty}^{\infty} P(y_1; y_2)\, dy_1$ are called the *marginal distributions* of $P(y_1; y_2)$. The marginal distributions, or projections, of a multi-variate Gaussian are also Gaussians.

Appendix D   Dirac's δ-function

Dirac's δ-function, $\delta(x)$, has been used several times in these notes. Heuristically, this function can be defined as one that is zero for all $|x| > 0$ and infinite for $x = 0$, in such a way that $\int_{-\infty}^{\infty} \delta(x)\,dx = 1$. It is surprising how far one can get with this definition, but it fails when it comes to discussing derivatives of the δ-function, for instance. In this appendix we attempt to discuss the δ-function in more detail, although finer mathematical aspects are left to the literature (for instance Lighthill (1964) and Berz (1967)).

From these introductory remarks it is evident that a δ-function only makes sense as an integrand, possibly multiplied by another function. A graph of $\delta(x)$ is, for instance, hardly meaningful. The notion of a δ-function can be made meaningful in terms of *generalized functions*. First we introduce the notion of an *acceptable function*, or a 'good function' in the sense of Lighthill (1964). An acceptable function is one that is everywhere differentiable any number of times and such that all its derivatives are $\mathcal{O}(|x|^{-N})$ as $|x| \to \infty$ for all N.

- **Example:** For instance $\exp(-x^2)$ is an acceptable (or 'good') function, whereas polynomials are not acceptable functions.

A sequence $f_n(x)$ of acceptable functions is called regular if, for any regular function $F(x)$, the limit $\lim_{n \to \infty} \int_{-\infty}^{\infty} f_n(x)F(x)\,dx$ exists. Two such regular sequences are *equivalent* if, for all acceptable functions $F(x)$, this limit is the same for each sequence.

- **Example:** Two sequences $\exp(-x^2/n^2)$ and $\exp(-x^4/n^4)$ are equivalent. For both of these two cases we have $\lim_{n \to \infty} \int_{-\infty}^{\infty} f_n(x)F(x)\,dx = \int_{-\infty}^{\infty} F(x)\,dx$.

A generalized function $f(x)$ is defined as a regular sequence $f_n(x)$ of acceptable functions, and two generalized functions are said to be equal if the corresponding regular sequences are equivalent. Thus, each generalized function is a class of all regular sequences equivalent to a given regular sequence. A generalized function $f(x)$ is said to be even or odd, respectively, if $\int_{-\infty}^{\infty} f(x)F(x)\,dx = 0$ for all odd or even acceptable functions $F(x)$.

With the notation introduced here, Dirac's $\delta(x)$-function can be defined by the sequence $f_n(x) = (n/\pi)^{1/2} \exp(-nx^2)$ and equivalent sequences, which have the property that $\lim_{n \to \infty} \int_{-\infty}^{\infty} f_n(x)F(x)\,dx = \int_{-\infty}^{\infty} \delta(x)F(x)\,dx = F(0)$. The δ-function is even in the sense discussed before.

One other representation for a δ-function is

$$\delta(x) = \lim_{L \to \infty} \frac{\sin(xL)}{\pi x}. \tag{D.1}$$

At $x = 0$ the value of the expression in (D.1) is L/π, whereas for all other x-values it oscillates with a period $2\pi/L$ that decreases as $L \to \infty$. The integral from $-\infty$ to ∞ of $\sin(xL)/(\pi x)$ is unity, independent of L. Therefore the limiting case of (D.1) has the necessary properties of a δ-function as $L \to \infty$. Use of (D.1) proves the often-used expression

$$\frac{1}{2\pi} \int_{-\infty}^{\infty} e^{ikx} dk = \delta(x),$$ (D.2)

since $\int_{-\infty}^{\infty} e^{ikx} dk$ can be considered as the limiting case of $\lim_{L\to\infty} \int_{-L}^{L} e^{ikx} dk$. We then have trivially

$$\frac{1}{2\pi} \int_{-\infty}^{\infty} \cos(kx)\, dk = \delta(x) \quad \text{while} \quad \int_{-\infty}^{\infty} \sin(kx)\, dk = 0.$$ (D.3)

As an alternative representation one might take

$$\delta(x) = \lim_{\alpha \to 0} \frac{1}{\pi} \frac{\alpha}{\alpha^2 + x^2}.$$ (D.4)

which again fulfills the requirements posed by the basic definition of a δ-function. Similarly we have

$$\delta(x) = \lim_{\alpha \to 0} \frac{1}{\alpha\sqrt{\pi}} e^{-(x/\alpha)^2},$$ (D.5)

Given a complete set of orthonormal functions, $\psi_n(x)$, a δ-function can be represented by

$$\delta(x - x') = \sum_{n=1}^{\infty} \psi_n(x')\psi_n^*(x),$$ (D.6)

if ψ_n corresponds to a discrete set, or, alternatively, if the basis functions belong to a continuous spectrum, then

$$\delta(x - x') = \int \psi_s(x')\psi_s^*(x)\, ds.$$ (D.7)

For comparison, we give a related expression for the Kronecker-delta as e.g.

$$\delta_{r,s+pN} = \frac{1}{N} \sum_{n=0}^{n=N-1} \exp[i2\pi(r-s)n/N],$$

where p is an integer.

The most important properties of δ-functions (Davydov, 1965, Pathria, 1996) can be summarized as

$$\int_\alpha^\beta F(x)\delta(x-a)\,dx = F(a) \qquad \text{if } \alpha < a < \beta, \tag{D.8}$$

$$f(x)\delta(x-a) = f(a)\delta(x-a), \tag{D.9}$$

$$\delta(x) = \delta(-x), \tag{D.10}$$

$$x\delta(x) = 0, \tag{D.11}$$

$$\delta(ax) = \frac{1}{|a|}\delta(x), \tag{D.12}$$

$$\int_\alpha^\beta \delta(a-x)\delta(x-b)\,dx = \delta(a-b) \qquad \text{if } \alpha < a, b < \beta, \tag{D.13}$$

$$\delta(x^2-a^2) = \frac{1}{2|a|}[\delta(x-a)+\delta(x+a)], \tag{D.14}$$

$$\delta(f(x)) = \sum_j \frac{\delta(x-x_j)}{\left|(df/dx)_{x=x_j}\right|}, \tag{D.15}$$

where in (D.15) the x_j are simple roots of $f(x)=0$. (D.14) is just a special, but often useful, version of (D.15). Note that the dimension of $\delta(x)$ is $[x]^{-1}$, where the dimension of the argument is $[x]$.

- **Example:** An important consequence of (D.11) is that the equation

$$xF(x) = G(x) \tag{D.16}$$

has not the solution $F(x) = G(x)/x$ as one might first expect, but rather $F(x) = G(x)/x + \lambda\delta(x)$, where λ is an arbitrary constant. Now, as already empha-sized, the presence of a δ-function is, strictly speaking, meaningful only under an integral sign. Such an integration is, however, problematic, unless $G(x=0)=0$. We can *prescribe* that the principal value of the integral has to be taken, and then λ can be considered as being a constant that compensates for the arbitrariness of this choice. The solution of (D.16) is consequently written as

$$F(x) = \mathcal{P}\frac{G(x)}{x} + \lambda\delta(x), \tag{D.17}$$

where $\mathcal{P}$ indicates that the principal value has to be taken upon integration. If, for some reason, it is known *a priori* that $F(x)$ is normalizable, then the normalizing condition $\int F(x)\,dx = 1$ can be used to determine λ, with $G(x)$ given, see e.g., van Kampen and Felderhof (1967).

The foregoing examples all assumed the δ-function to depend on one variable only. A 'three-dimensional' version can be defined as $\delta(\mathbf{r})$, which can be interpreted as $\delta(\mathbf{r}) = \delta(x)\delta(y)\delta(z) = (2\pi)^{-3}\int \exp(i\mathbf{k}\cdot\mathbf{r})\,dk_x\,dk_y\,dk_z$, having the basic property

$$\int \delta(\mathbf{r}-\mathbf{r}_0)F(\mathbf{r})\,dx\,dy\,dz = F(\mathbf{r}_0).$$

In particular, the dimension of $\delta(\mathbf{r})$ is length^{-3}.

The following are useful relations, which have no counterpart in the one-dimensional version:

$$\delta(\mathbf{r}) = \frac{1}{2\pi r^2}\delta(r),$$

$$\delta(\mathbf{r}' - \mathbf{r}) = \frac{2}{r^2}\delta(\mathbf{n}' - \mathbf{n})\delta(r' - r),$$

where, in the latter relation, $\mathbf{n}'$ and $\mathbf{n}$ are unit vectors along $\mathbf{r}'$ and $\mathbf{r}$, respectively. An integration over r should start at $r = 0$.

One important consequence of introducing a δ-function is that continuous and discrete charge distributions can be treated in formally the same way. In Poisson's equation, for instance, $\nabla \cdot \mathbf{D} = \rho(\mathbf{r}, t)$, a discrete charge distribution can be represented as $\sum_j q_j \delta(\mathbf{r} - \mathbf{R}_j(t))$, where $\mathbf{R}_j(t)$ are the positions, at time t, of charges q_j.

- **Example:** The equation

$$\nabla \cdot \mathbf{E} = \frac{a}{4\pi}\delta(\mathbf{r})$$

  has the solution $\mathbf{E} = a\hat{\mathbf{r}}/r^2$ for all $|\mathbf{r}| > 0$, since, for this case with spherical symmetry, we have $\nabla \cdot (\hat{\mathbf{r}}/r^2) = r^{-2}\,\partial(r^2\hat{\mathbf{r}}/r^2)/\partial r = 0$ for all $|\mathbf{r}| > 0$. Using Gauss's divergence theorem $\int \nabla \cdot \mathbf{E}\, dx\, dy\, dz = \oint \mathbf{E} \cdot d\mathbf{s}$ on a spherical surface around the origin, we find $\int \nabla \cdot \hat{\mathbf{r}}/r^2\, dx\, dy\, dz = 4\pi$, completing the proof.

D.1   Derivatives of δ-functions

The derivative of a δ-function can be defined from (D.1) as the limiting case of

$$\delta'(x) = \frac{1}{\pi}\lim_{L\to\infty}\left(\frac{L\cos(xL)}{x} - \frac{\sin(xL)}{x^2}\right). \tag{D.18}$$

We find that

$$\int_\alpha^\beta \delta'(x)F(x)\, dx = -F'(0),$$

with $\alpha < 0 < \beta$. More generally, the nth derivative satisfies $\int_\alpha^\beta \delta^{(n)}(x)F(x)\, dx = (-1)^n F^{(n)}(0)$. In particular, the first derivative of the δ-function satisfies the relation

$$x\delta'(x) = -\delta(x).$$

The derivative of the δ-function is an odd function in the sense discussed before.

D.2   Functions related to δ-functions

We can introduce other functions that are related to the δ-function, such as

$$\delta_+(x) = \delta_-^*(x) = \frac{1}{i2\pi}\lim_{\alpha\to 0}\frac{1}{x - i\alpha}. \tag{D.19}$$

Using (D.19) and (D.4), we find

$$\delta_+(x) + \delta_-(x) = \lim_{\alpha \to 0} \frac{1}{\pi} \frac{\alpha}{\alpha^2 + x^2} = \delta(x), \tag{D.20}$$

$$\delta_+(x) - \delta_-(x) = \lim_{\alpha \to 0} \frac{1}{i\pi} \frac{x}{\alpha^2 + x^2} = \mathcal{P} \frac{1}{i\pi x}, \tag{D.21}$$

where $\mathcal{P}$ indicates that the principal value should be taken upon integration with respect to x. The second part of the relation (D.21) can be understood by noting that $x^2/(x^2 + \alpha^2)$ weights the two sides of a multiplying function equally with respect to the point $x = 0$.

Sometimes one may encounter a function

$$\xi(x) = \lim_{k \to \infty} \frac{1 - e^{-ikx}}{ix} \tag{D.22}$$

as a multiplier to an integrand. Such integrals can be evaluated by noting that

$$\xi(x) = \pi \delta(x) - i\mathcal{P} \frac{1}{x}. \tag{D.23}$$

The function

$$\lim_{\alpha \to 0} \frac{1}{x - i\alpha} = i\pi \delta(x) + \mathcal{P} \frac{1}{x} \tag{D.24}$$

has the same property and relates $\xi(x)$ to $\delta_\pm(x)$. The latter example serves to illustrate the argument about selecting the principal value when integrating with respect to x; an integral containing the left-hand side of (D.24) will, for all $\alpha \neq 0$, have a pole in the complex plane at $z = \{x, \alpha\}$. As $\alpha \to 0$ this pole moves toward the real x-axis, but at all times the path of integration goes *under* the pole. In order to retain the integral as an analytic function of α also in the limit $\alpha = 0$, we take the path of integration to be locally a small half-circle with radius r around the pole in this limit, and subsequently let $r \to 0$. This procedure produces the principal value of the integral and also half the residue which is accounted for by $i\pi \delta(x)$ in (D.24).

The δ-function can be extended to be a function of a complex variable z. Using (D.4) we find that $\delta(z)$ has two simple poles at the points ix and $-ix$, with residues equal to $1/(2\pi i)$ and $-1/(2\pi i)$, respectively (Davydov, 1965). When integrating expressions containing $\delta(z)$ one must make certain that the integration path goes between these two poles. The relations (D.19)–(D.21) remain valid also for complex variables. In those cases we have

$$\delta_-(z) = \delta_+(-z) = \delta_+^*(z) = [\delta_+(z^*)]^*.$$

One can also write

$$\delta_+(z) = \frac{1}{2\pi i z}, \qquad \delta_-(z) = -\frac{1}{2\pi i z},$$

recalling again that the path of integration has to run above and below the point $z = 0$ for the two cases, respectively.

D.2.1   Heaviside's step function

Heaviside's unit step function, $H(t) = 1$ for $t > 0$ and $H(t) = 0$ for $t < 0$, can be obtained from the δ-function by

$$H(t) = \int_{-\infty}^{t} \delta(\tau) \, d\tau.$$

The function $\text{sign}(t > 0) = 1$ and $\text{sign}(t < 0) = -1$ is closely related to Heaviside's step function. Evidently, we have $\text{sign}(t) = 2H(t) - 1 = H(t) - H(-t)$. By differentiation we find $dH(t)/dt = \delta(t)$ and $d\text{sign}(t)/dt = 2\delta(t)$.

Appendix E   Physical constants

Speed of light in vacuum	$c = 2.9979 \times 10^8$ m s^{-1}
Permittivity of free space	$\varepsilon_0 = 8.8542 \times 10^{-12}$ F m^{-1}
Permeability of free space	$\mu_0 = 4\pi \times 10^{-7}$ H m^{-1}
Planck's constant	$h = 6.6256 \times 10^{-34}$ J s
	$\hbar \equiv h/(2\pi) = 1.0546 \times 10^{-37}$ J s
Gravitational constant	$G = 6.6726 \times 10^{-11}$ N m^2 kg^{-2}
Standard gravitational acceleration	$g_n = 9.80665$ m s^{-2}
Elementary charge	$e = 1.6022 \times 10^{-19}$ C
Electron mass	$m = 9.1095 \times 10^{-31}$ kg
Proton mass	$M = 1.6726 \times 10^{-27}$ kg
Atomic mass unit (^{12}C/12)	amu $= 1.6604 \times 10^{-27}$ kg
Bohr radius, $4\pi\varepsilon_0\hbar^2/(e^2 m)$	$a_0 = 5.2918 \times 10^{-11}$ m
Atomic cross-section	$\pi a_0^2 = 8.7974 \times 10^{-21}$ m^2
Classical electron radius	$e^2/(4\pi\varepsilon_0 mc^2) = 2.82 \times 10^{-15}$ m
Bohr magneton	$\mu_B \equiv \hbar e/(2m) = 9.27 \times 10^{-24}$ A m^2
Nuclear magneton	$\mu_N \equiv \hbar e/(2M) = 5.05 \times 10^{-27}$ A m^2
Fine structure constant	$\alpha \equiv e^2/2\varepsilon_0 hc \approx 1/137$
Boltzmann constant	$\kappa = 1.3807 \times 10^{-23}$ J K^{-1}
Avogadro number (molecules/mol)	$N_A = 6.0220 \times 10^{23}$
Standard pressure (760 Torr = 1 atm)	$p_0 = 1.0133 \times 10^5$ Pa
Ideal-gas molar volume	$V_{mol} = 0.022414$ m^3 mol^{-1}
Loschmidt number (ideal gas density at STP)	$n_L = 2.6868 \times 10^{25}$ m^{-3}
Gas law constant	$R = 8.31$ J mol^{-1} K^{-1}
Absolute zero temperature	$T_0 = -273.15$ °C
Conversion eV $\rightarrow$ J	1 eV $= 1.602 \times 10^{-19}$ J
Conversion cal $\rightarrow$ J	1 cal $= 4.186$ J
Conversion kW h $\rightarrow$ J	1 kW h $= 3.6 \times 10^6$ J
Conversion erg $\rightarrow$ J	1 erg $= 10^{-7}$ J
Conversion Btu $\rightarrow$ J	1 Btu $= 1.054 \times 10^3$ J
Atmospheric pressure $\rightarrow$ Pa	1 atm $= 1.01 \times 10^5$ Pa
mm mercury $\rightarrow$ Pa	1 mmHg $= 133.322$ Pa
mm mercury $\rightarrow$ kg m^{-2}	1 mmHg $= 13.60$ kg m^{-2}
mm mercury $\rightarrow$ atm	1 mmHg $= 1.316 \times 10^{-3}$ atm

Appendix F   The MKSA units

This book uses MKSA units; meters (m) for length, kilograms (kg) for mass, seconds (s) for time, and amperes (A) for current. There is an additional, or supplementary unit for temperature, kelvins (K). At times it is argued that angles are a supplementary unit also, but these can be obtained as a length by measuring the length along the arc of a unit circle giving $180° \rightarrow \pi$ m, from which usually the 'm' is omitted, and 'radians' are preferred without reference to the circle of unit radius. Luminous intensity is measured in candelas (cd). If you stay within the MKSA system the result of a correct calculation will always have the correct physical dimension. In reality we do not always trust our ability to do correct calculations and it is wise to check, from time to time, whether results are dimensionally correct. In particular, it was advocated in Section 4.3 that in some cases a result can be obtained by dimensional arguments *alone*. In such cases one has to express other units in MKSA units, for instance charge which is measured in coulombs (C). Now, this is easy, since current is charge per second by definition, giving C = A s. In other cases the calculation might take a little more labor, so, to save the reader some time, the following table is given, in terms of farads (F) for capacitance, henries (H) for inductance, joules (J) for energy, newtons (N) for force, ohms (Ω) for resistance, siemens (S) for conductance, pascals (Pa) for pressure, teslas (T) for magnetic flux density, volts (V) for voltage, watts (W) for power, and webers (Wb) for magnetic flux. The notation [a] will here denote the dimension of the quantity *a*.

Quantity	Unit	MKSA unit
Boltzmann's constant	$J\,K^{-1}$	$kg\,m^2\,K^{-1}\,s^{-2}$
Capacitance	$F = A\,s\,V^{-1}$	$A^2\,s^4\,kg^{-1}\,m^{-2}$
Charge	C	$A\,s$
Conductance	$S = \Omega^{-1}$	$A^2\,s^3\,kg^{-1}\,m^{-2}$
Conductivity	$S\,m^{-1}$	$A^2\,s^3\,kg^{-1}\,m^{-3}$
Electric field	$V\,m^{-1}$	$kg\,m\,A^{-1}\,s^{-3}$
Energy	$J = N\,m$	$kg\,m^2\,s^{-2}$
Force	N	$kg\,m\,s^{-2}$
Gravitational constant, $[G]$	$N\,m^2\,kg^{-2}$	$m^3\,kg^{-1}\,s^{-2}$
Heat capacity	$J\,kg^{-1}\,K^{-1}$	$m^2\,s^{-2}\,K^{-1}$
Induction	$H = V\,s\,A^{-1}$	$kg\,m^2\,A^{-2}\,s^{-2}$
Magnetic flux	$Wb = V\,s$	$kg\,m^2\,A^{-1}\,s^{-2}$
Magnetic flux density	$T = Wb\,m^{-2}$	$kg\,A^{-1}\,s^{-2}$
Permeability of free space, $[\mu_0]$	$H\,m^{-1}$	$kg\,m\,A^{-2}\,s^{-2}$
Permittivity of free space, $[\varepsilon_0]$	$F\,m^{-1}$	$A^2\,s^4\,kg^{-1}\,m^{-3}$
Planck's constant, $[h]$	$J\,s$	$kg\,m^2\,s^{-1}$
Power	$W = J\,s^{-1} = V\,A$	$kg\,m^2\,s^{-3}$
Pressure	$Pa = N\,m^{-2}$	$kg\,m^{-1}\,s^{-2}$
Resistance	$\Omega = V\,A^{-1}$	$kg\,m^2\,A^{-2}\,s^{-3}$
Resistivity	$\Omega\,m$	$kg\,m^{-3}\,A^{-2}\,s^{-3}$
Viscosity (dynamic)		$kg\,m^{-1}\,s^{-1}$
Viscosity (kinematic)		$m^2\,s^{-1}$
Voltage	$V = W\,A^{-1} = J\,C^{-1}$	$kg\,m^2\,A^{-1}\,s^{-3}$

References

Abramowitz, M. and **Stegun, A. I.** (1972) *Handbook of Mathematical Functions*, Dover, New York.

Alkemade, C. T. J. (1958) *Physica*, Vol. 24, 1029.

Bak, P. (1997) *How Nature Works*, Oxford University Press, London.

Bak, P., Tang, C., and **Wiesenfeld, K.** (1987) *Phys. Rev. Lett.*, Vol. 59, 381.

Balescu, R. (1997) *Statistical Dynamics*, Imperical College Press, London.

Bayes, T. (1763) *Phil. Trans.*, Vol. 53, 370.

Bekefi, G. (1966) *Radiation Processes in Plasmas*, John Wiley & Sons, New York.

Bendat, J. S. (1958) *Principles and Applications of Random Noise Theory*, John Wiley & Sons, New York.

Bendat, J. S. and **Piersol, A. G.** (1986) *Random Data, Analysis and Measurement Procedures*, 2nd ed., John Wiley & Sons, New York.

Bertrand, J. (1889) *Calcul des probabilités*, Gauthier-Villars, Paris.

Berz, E. (1967) *Verallgemeinerte Funktionen und Operatoren*, Hochshultaschenbücher 122/122a, Bibliographisches Institut, Mannheim.

Bode, H. W. (1956) *Network Analysis and Feedback Amplifier Design*, D. Van Nostrand, Princeton, New Jersey.

Brigham, E. O. (1974). *The Fast Fourier Transform*, Prentice-Hall, New Jersey.

Brown, R. (1827) *A Brief Account of Microscopical Observations Made in the Months of June, July, and August, 1827, on the Particles Contained in the Pollen Plants; and on the General Existence of Active Molecules in Organic and Inorganic Bodies*. Private printing.

Brown, R. (1828) *Phil. Mag.*, Vol. 4, 161.

Brown, R. (1828) *Ann. Phys. Chem.*, Vol. 14, 294.

Buckingham, E. (1914) *Phys. Rev.*, Vol. 4, 345.

Bunkin, F. V. (1957) *Sov. Phys. JETP*, Vol. 5, 227 and *ibid.* 665.

Callen, H. B. and **Welton, T. A.** (1951) *Phys. Rev.*, Vol. 83, 34.

Campbell, N. R. (1909) *Proc. Phil. Soc.*, Vol. 15, 117 and *ibid.* 310.

Carslaw, H. S. and **Jaeger, J. C.** (1959) *Conduction of Heat in Solids*, Oxford University Press, London.

Champeney, D. C. (1973) *Fourier Transforms and their Physical Applications*, Academic Press, London.

Chandrasekhar, S. (1943) *Rev. Mod. Phys.*, Vol. 15, 1. Reprinted in: (1954) *Selected Papers on Noise and Stochastic Processes* (Ed. Wax, N.) Dover, New York, pp. 3–91.

Chen, F. F. (1983) *Introduction to Plasma Physics and Controlled Fusion*, Plenum Press, New York.

Chung, K. L. and **Fuchs, W. H. J.** (1951) *Mem. Amer. Math. Soc.*, No. 6.

Clarke, J. and **Voss, R. F.** (1974) *Phys. Rev. Lett.*, Vol. 33, 24.

Clemmow, P. C. and **Dougherty, J. P.** (1990) *Electrodynamics of Particles and Plasmas*, Addison-Wesley, Redwood City, California.

Cox, D. R. and **Miller, H. D.** (1965) *The Theory of Stochastic Processes*, Chapman and Hall, London.

Davenport, W. B. and **Root, W. L.** (1958) *An Introduction to the Theory of Random Signals and Noise*, McGraw-Hill, New York.

Davydov, A. S. (1965) *Quantum Mechanics*, Pergamon Press, Oxford.

Deutsch, D. H. (1992) *Nature*, Vol. 357, 354.

Dreïzin, Yu. A. and **Dykhne, A. M.** (1972) *Zh. Éksp. Teor. Fiz.*, Vol. 63, 242 [(1973) *Sov. Phys. JETP*, Vol. 36, 127].

Einstein, A. (1956) Reprint of translations of the original papers *On the movement of small* *particles suspended in a stationary liquid demanded by the molecular-kinetic theory of heat, On the theory of the Brownian movement, A new determination of molecular dimensions, Theoretical observations of the Brownian motion,* and *Elementary theory of the Brownian motion* (Ed. Fürth, R., transl. Cowper, A. D.) Dover, New York.

Erdélyi, A. (1956) *Asymptotic Expansions*, Dover, New York.

Feller, W. (1968) *An Introduction to Probability Theory and its Applications*, Vol. I, John Wiley & Sons, New York.

Feller, W. (1971) *An Introduction to Probability Theory and its Applications*, Vol. II, John Wiley & Sons, New York.

Fermi, E. (1949) *Phys. Rev.*, Vol. 75, 1169.

Feynman, R. P., Leighton, R. B., and Sands, M. (1963) *The Feynman Lectures on Physics*, Addison-Wesley, Reading, Massachusetts.

Fried, B. D. and Conte, S. D. (1961) *Plasma Dispersion Function*, Academic Press, New York.

Furutsu, K. (1963) *J. Res. Nat. Bur. Standards* D, Vol. 67, 303.

Gerlach, W. (1927) *Naturwissenschaft*, Vol. 15, 15.

Gibbings, J. C. (1982) *J. Phys. A: Math. Gen.*, Vol. 15, 1991.

Greene, J. M. (1979) *J. Math. Phys.*, Vol. 20, 1183.

Hammersley, J. M. (1961) In: *Proceedings of the Fourth Berkeley Symposium on Mathematics, Statistics and Probability*, Vol. 3 (Ed. Neyman, J.) University of California Press, Berkeley.

Hasselman, K., Munk, W., and MacDonald, G. (1963) In: *Time Series Analysis* (Ed. Rosenblatt, M.) Wiley, New York, p. 125.

Hauge, E. H. and Martin-Löf, A. (1973) *J. Statist. Phys.*, Vol. 7, 259.

Haus, H. A. (1961) *J. Appl. Phys.*, Vol. 32, 493.

Hogg, R. V. and Craig, A. T. (1970) *Introduction to Mathematical Statistics*, 3rd ed., Macmillan, London.

Horowitz, P. and Hill, W. (1980) *The Art of Electronics*, Cambridge University Press, Cambridge.

Hubbard, J. (1961) *Proc. Roy. Soc.* A, Vol. 260, 114.

Jackson, E. A. (1968) *Equilibrium Statistical Mechanics*, Prentice-Hall, New Jersey.

Johnson, J. B. (1925) *Phys. Rev.*, Vol. 26, 71.

Johnson, J. B. (1928) *Phys. Rev.*, Vol. 32, 97.

Kac, M. (1946) *Am. Math. Monthly*, Vol. 54, 369. Reprinted in: (1954) *Selected Papers on Noise and Stochastic Processes* (Ed. Wax, N.) Dover, New York, pp. 295–317.

Kampen, N. G. van (1960) *Physica*, Vol. 26, 585.

Kampen, N. G. van (1961) *J. Math. Phys.*, Vol. 2, 592.

Kampen, N. G. van (1963) *J. Math. Phys.*, Vol. 4, 190.

Kampen, N. G. van (1981) *Stochastic Processes in Physics and Chemistry*, North-Holland, Amsterdam.

Kampen, N. G. van and Felderhof, B. U. (1967) *Theoretical Methods in Plasma Physics*, North-Holland, Amsterdam.

Kendall, D. G. (1961) In: *Proceedings of the Fourth Berkeley Symposium on Mathematics, Statistics and Probability*, Vol. 3 (Ed. Neyman, J.) University of California Press, Berkeley.

Khinchine, A. I. (1949) *Mathematical Foundations of Statistical Mechanics*, Dover, New York.

Kim, Y. C. and Powers, E. J. (1979) *IEEE Trans. Plasma Sci.*, Vol. 7, 120.

Kittel, C. (1968) *Introduction to Solid State Physics*, 3rd ed., John Wiley & Sons, New York.

Klafter, J., Schlesinger, M. F., and Zumofen, G. (1996) *Phys. Today*, Vol. 49, No. 2, 33.

Kogan, Sh. (1996) *Electronic Noise and Fluctuations in Solids*, Cambridge University Press, Cambridge.

Kubo, R. (1957) *J. Phys. Soc. Japan*, Vol. 12, 570.

Kubo, R. and Tomita, K. (1954) *J. Phys. Soc. Japan*, Vol. 9, 888.

Landau, L. D. and Lifshitz, E. M. (1960) *Electrodynamics of Continuous Media*, Pergamon Press, Oxford.

Lax, M. (1960) *Rev. Mod. Phys.*, Vol. 32, 25.

Lee, J. F., Sears, F. W., and Turcotte, D. L. (1973) *Statistical Thermodynamics*, 2nd ed., Addison-Wesley, New York.

Lee, P. M. (1989) *Bayesian Statistics: An Introduction*, Halsted Press, New York.

Leontovich, M. A. and **Rytov, S. M.** (1952) *Zh. Éksp. Teor. Fiz.*, Vol. 23, 246.

Lévi, P. (1937) *Théorie de l'addition des variables aléatoires*, Gauthier-Villiers, Paris.

Lichtenberg, A. J., Lieberman, M. A., and **Cohen, R. H.** (1980) *Physica* D, Vol. 1, 291.

Lieberman, M. A. and **Lichtenberg, A. J.** (1972) *Phys. Rev.* A, Vol. 5, 1852.

Lighthill, M. J. (1964) *An Introduction to Fourier Analysis and Generalized Functions*, Cambridge University Press, Cambridge.

Lii, K. S., Rosenblatt, M., and **Van Atta, C.** (1976) *J. Fluid Mech.*, Vol. 77, 45.

Lorentz, H. A. (1921) *Lessen over theoretische natuurkunde.* Vol. 5 *Kinetische problemen.* E. J. Brill, Leiden.

MacDonald, D. K. C. (1962) *Noise and Fluctuations: An introduction.* John Wiley & Sons, New York.

Martin, P. C. (1968) In: *Problème à N corps. Many-Body Physics* (Ed. DeWitt, C. and Balian, R.) Gordon and Breach, New York, pp. 39–136.

Mercier, R. (1964) *Proc. Phys. Soc.*, Vol. 83, 819.

Moore, W. J. (1974) *J. Appl. Phys.*, Vol. 45, 1896.

Morse, P. M. (1969) *Thermal Physics*, 2nd ed., W. A. Benjamin, New York.

Musha, T. and **Higuchi, H.** (1976) *Jap. J. Appl. Phys.*, Vol. 15, 1271.

Narasimhan, T. N. (1999) *Rev. Geophys.*, Vol. 37, 151.

Nägeli, von (1879) *Münch. Sitzungsber. Math. Phys.*, Vol. 9, 389.

Novikov, E. A. (1964) *Zh. Éksp. Teor. Fiz.*, Vol. 47, 1919.

Nyquist, H. (1928) *Phys. Rev.*, Vol. 32, 110.

Oppenheim, A. V. and **Schafer, R. W.** (1975) *Digital Signal Processing*, McGraw-Hill, New York.

Papoulis, A. (1991) *Probability, Random Variables and Stochastic Processes*, 3rd ed., McGraw-Hill, New York.

Pathria, R. K. (1996) *Statistical Mechanics*, 2nd ed., Butterworth-Heinemann, Oxford.

Pécseli, H. L. (1976) *J. Appl. Phys.*, Vol. 47, 2415.

Pécseli, H. L. and **Trulsen, J.** (1993) *Plasma Phys. Contr. Fusion*, Vol. 35, 1701.

Perrin, J. (1916) *Atoms*, Constable, London.

Pustylnikov, L. D. (1978) *Trans. Moscow Math. Soc.*, Vol. 2, 1.

Rayleigh, see **Strutt, J. W.**

Reichl, L. E. (1998) *A Modern Course in Statistical Physics*, 2nd ed., John Wiley & Sons, New York.

Rice, S. O. (1944) *Bell Syst. Tech. J.*, Vol. 23, 282–332 and *ibid.* (1945) Vol. 24, 46–156; most readily available in: (1954) *Selected Papers on Noise and Stochastic Processes* (Ed. Wax, N.) Dover, New York, pp. 133–294.

Rostoker, N. (1961) *Nucl. Fusion*, Vol. 1, 101.

Rowland, E. N. (1936) *Proc. Cambridge Phil. Soc.*, Vol. 32, 580.

Schick, K. L. and **Verveen, A. A.** (1974) *Nature*, Vol. 251, 599.

Schottky, W. (1918) *Ann. Phys. Leipzig*, Vol. 57, 541.

Schottky, W. (1926) *Phys. Rev.*, Vol. 28, 74.

Solomon, T., Weeks, E., and **Swinney, H.** (1993) *Phys. Rev. Lett.*, Vol. 71, 3975.

Strutt, J. W. (Baron Rayleigh) (1902) *Scientific Papers*, Vol. III, Cambridge University Press, Cambridge.

Sveshnikov, A. A. (1978) *Problems in Probability Theory, Mathematical Statistics and Theory of Random Functions*, Dover, New York.

Symon, K. R. (1960) *Mechanics*, Addison-Wesley, Reading, Massachusetts.

Takács, L. (1960) *Stochastic Processes, Problems and Solutions*, Methuen, London.

Taylor, J. B. (1960) *Phys. Fluids*, Vol. 3, 792.

Thompson, W. B. and **Hubbard, J.** (1960) *Rev. Mod. Phys.*, Vol. 32, 714.

Titchmarsh, E. C. (1950) *The Theory of Functions*, 2nd ed., Oxford University Press, London.

Tolman, R. C. (1938) *The Principles of Statistical Mechanics*, Oxford University Press, London.

Uhlenbeck, G. E. and **Goudsmit, S.** (1929) *Phys. Rev.*, Vol. 34, 145.

Uhlenbeck, G. E. and **Ornstein, L. S.** (1930) *Phys. Rev.*, Vol. 36, 823. Reprinted in: (1954) *Selected Papers on Noise and Stochastic Processes* (Ed. Wax, N.) Dover, New York, pp. 93–111.

Ulam, S. (1961) In: *Proceedings of the Fourth Berkeley Symposium on Mathematics,*

Statistics and Probability, Vol. 3 (Ed. Neyman, J.) University of California Press, Berkeley.

Wang, M. C. and **Uhlenbeck, G. E.** (1945) *Rev. Mod. Phys.*, Vol. 17, 323. Reprinted in: (1954) *Selected Papers on Noise and Stochastic Processes* (Ed. Wax, N.) Dover, New York, pp. 113–132.

Yaglom, A. M. (1962) *An Introduction to the Theory of Stationary Random Functions*, Prentice-Hall, New York (also available from Dover, New York, 1973).

Yeh, K. C. and **Liu, C. H.** (1972) *Theory of Ionospheric Waves*, Academic Press, New York.

Zaslavskii, G. M. and **Chirikov, B. V.** (1964) *Dokl. Akad. Nauk SSSR*, Vol. 159, 306 (*Sov. Phys. Doklady*, Vol. 9, 989, 1965).

Zernike, F. (1929) In: *Handbuch der Physik*, Vol. 3 (Ed. Geiger, R. and Scheel, K. C.) Springer, Berlin.

Index

1/f noise, 123

absorbing barrier, 91, 147
acceptable function, 173
active media, 47
adiabatic processes, 162
almost-stationary process, 18
ammeter, 77
anisotropic media, 49, 62
Archimedes' law, 144
asymptotic expansion, 28, 96
auto-correlation, 14, 15, 21, 116, 122

barriers, 90
Bayes' rule, 10, 108, 129, 138
Bertrand's example, 2
binomial distribution, 98, 167
bispectrum, 120
bivariate Fokker–Planck equation, 133
black-body radiation, 51
blue light, 99
Bochner's theorem, 25
Bode relations, 56
box function, 15, 25, 113
Brownian motion, 69, 73, 95, 146
Brownian-motion process, 134

Campbell's theorem, 105, 110
capacitor plates, 30, 153
carbon resistors, 124
causality, 54
central-limit theorem, 115
Čerenkov radiation, 62, 133
change of variable, 8
Chapman–Kolmogorov equation, 130
characteristic function, 7, 9, 112
characteristic timescales, 17
chi-squared probability density, 8
circuit impedance, 36
coagulation in colloids, 146
coherence function, 12
coin-tossing game, 84

collision frequency, 34, 39, 170
collisional damping, 39
collisionless damping of plasma waves, 60
collisions, 33, 169
colloids, 146
complex variables, 10, 12
condensation, 101
conditional probability, 10, 129, 156
convective time derivative, 141
correlation coefficient, 10
correlation function, 11, 13, 116, 137
correlation matrix, 12, 19, 137, 172
correlation time, 17
covariance, 10
critical point, 102
cross-correlation, 14, 118
cross-phase spectrum, 28
cross-power spectrum, 21, 28
cross-spectral density, 21
cumulants, 109
cumulative distribution function, 6

D'Arsonval galvanometer, 77
dc value of a signal, 24
density fluctuations, 98
dichotomous Markov process, 135
dielectric function, 47, 55
diffusion coefficient, 70, 74, 93, 95, 134
diffusion equation, 70, 93, 95, 134
dimensional arguments, 46
diode, 44, 121, 152, 155
Dirac's δ-function, 6, 24, 39, 47, 57, 70, 144, 173
discretized signals, 118
dispersion relation, 49, 51, 59, 61
displacement current, 30, 37, 61
distribution function, 6
dressed particles, 62
Drude–Lorentz model, 67
drunkard's walk, 84

Edgeworth series, 115
Einstein relation, 71, 75, 143
elastic collisions, 72

187